Fifth Generation Warfare

This book outlines the concept of fifth generation warfare (5GW) and demonstrates its relevance for understanding contemporary conflicts.

Non-kinetic modes of attack and war waged by groups or nonstate actors at the societal level have been termed 5GW. This book discusses the theory of generational warfare and explores the key ideas of 5GW, such as secrecy, the manipulation of proxies, the manipulation of identity and culture (including disinformation and big data), and the use of psychological warfare. These techniques are used to achieve strategic objectives, such as inducing desired behavior and controlling human terrain, without resorting to overt war or overt violence. The text expands the debate on 5GW by exploring emerging technologies and how they could be used to maliciously shape human society and even to maliciously change the genetic makeup of a population for the purpose of unprecedented social control. The work closes with comments on the possibility of a sixth generation of warfare, which could target technical systems to possibly cause a society to collapse through strategic sabotage. Overall, the book demonstrates the relevance of 5GW for understanding contemporary conflicts, from the Arab Spring to the war in Ukraine, in terms of the need to dominate the human domain.

This book will be of interest to students of security and technology, defense studies, and international relations.

Armin Krishnan is an associate professor and director of security studies at East Carolina University, U.S.A., where he teaches U.S. foreign policy, international security, and intelligence studies. He is the author of several books, including *Military Neuroscience and the Coming Age of Neurowarfare* (2016).

Routledge Studies in Conflict, Security and Technology

Series Editors: Mark Lacy, *Lancaster University*, Dan Prince, *Lancaster University*, and Sean Lawson, *University of Utah*

The *Routledge Studies in Conflict, Technology and Security* series aims to publish challenging studies that map the terrain of technology and security from a range of disciplinary perspectives, offering critical perspectives on the issues that concern publics, business and policymakers in a time of rapid and disruptive technological change.

Militarising Artificial Intelligence
Theory, Technology and Regulation
Nik Hynek and Anzhelika Solovyeva

Understanding the Military Design Movement
War, Change and Innovation
Ben Zweibelson

Artificial Intelligence and International Conflict in Cyberspace
Edited by Fabio Cristiano, Dennis Broeders, François Delerue, Frédérick Douzet, and Aude Géry

Digital International Relations
Technology, Agency, and Order
Edited by Corneliu Bjola & Markus Kornprobst

Theorising Future Conflict
War Out to 2049
Mark Lacy

Fifth Generation Warfare
Dominating the Human Domain
Armin Krishnan

For more information about this series, please visit: https://www.routledge.com/Routledge-Studies-in-Conflict-Security-and-Technology/book-series/CST

Fifth Generation Warfare

Dominating the Human Domain

Armin Krishnan

Routledge
Taylor & Francis Group
LONDON AND NEW YORK

First published 2024
by Routledge
4 Park Square, Milton Park, Abingdon, Oxon OX14 4RN

and by Routledge
605 Third Avenue, New York, NY 10158

Routledge is an imprint of the Taylor & Francis Group, an informa business

© 2024 Armin Krishnan

British Library Cataloguing-in-Publication Data
A catalogue record for this book is available from the British Library

Library of Congress Cataloging-in-Publication Data
Names: Krishnan, Armin, 1975– author.
Title: Fifth Generation Warfare : dominating the human domain / Armin
    Krishnan.
Other titles: 5GW
Description: Abingdon, Oxon ; New York, NY : Routledge, 2024. |
    Series: Routledge studies in conflict, security and technology |
    Includes bibliographical references and index.
Identifiers: LCCN 2023042210 (print) | LCCN 2023042211 (ebook) |
    ISBN 9781032501192 (hbk) | ISBN 9781032501208 (pbk) |
    ISBN 9781003396963 (ebk)
Subjects: LCSH: Hybrid warfare. | Psychological warfare. | Information
    warfare. | Disinformation. | Cyberspace operations (Military science) |
    Military art and science—Technological innovations.
Classification: LCC U167.5.I8 K75 2024 (print) | LCC U167.5.I8 (ebook) |
    DDC 355.02—dc23/eng/20231212
LC record available at https://lccn.loc.gov/2023042210
LC ebook record available at https://lccn.loc.gov/2023042211

ISBN: 978-1-032-50119-2 (hbk)
ISBN: 978-1-032-50120-8 (pbk)
ISBN: 978-1-003-39696-3 (ebk)

DOI: 10.4324/9781003396963

Typeset in Times New Roman
by Apex CoVantage, LLC

In Memory of My Parents

Contents

List of Figures   *viii*
List of Tables   *ix*
List of Abbreviations   *x*

Introduction   1

1   Fourth Generation Warfare   11

2   Beyond the Fourth Generation   38

3   The Counterinsurgency Paradigm in 5GW   68

4   Cultural and Cognitive Warfare   98

5   Nano-Info-Bio-Cogno Technologies   128

6   Towards the Sixth Generation of Warfare   160

Index   *194*

Figures

2.1 OODA Loop   48
3.1 Domains of War   70
6.1 Critical Infrastructure   165

Tables

2.1	The Seven Gradients of War	51
4.1	The Stalinist Purges as 5GW	102
4.2	The Arab Spring as 5GW	107
4.3	The Havana Syndrome as 5GW	110
4.4	Manipulation of the COVID-19 Crisis as 5GW	115

Abbreviations

AI	Artificial intelligence
BCI	Brain–computer interface
BWC	Biological Weapons Convention
CBDC	Central bank digital currency
CCP	Chinese Communist Party
CRISPR	Clustered regularly interspersed short palindromic repeats
DARPA	Defense Advanced Research Projects Agency
DDOS	Distributed denial of service attack
DEW	Directed energy weapon
EEG	Electroencephalography
EMP	Electromagnetic pulse
HEMP	High-altitude electromagnetic pulse
HPM	High-power microwaves
HW	Hybrid warfare
ISR	Intelligence, surveillance, and reconnaissance
LPT	Large power transformer
MRI	Magnetic resonance imaging
NBIC	Nano-bio-info-cogno technologies
NGO	Non-governmental organization
NLW	Non-lethal weapons
NNEMP	Non-nuclear electromagnetic pulse
NPP	New physical principles weapons
OODA	Observe–orient–decide–act loop
PLA	People's Liberation Army
PSYACT	Psychological action
PSYOP	Psychological operation
RF	Radiofrequency
RFID	Radiofrequency identification
RMA	Revolution in military affairs
SOF	Special Operations Forces
SWS	Sentient World Simulation

Introduction

In today's world, the utility of force and major war is much diminished. War has become too costly and too destructive to function as a reasonable mechanism for settling disputes between sovereign states. War between major nuclear-weapon states could literally usher in the end of humanity.[1] Even conventional war among great powers on the scale of the Second World War would be, with current technology, extremely destructive and end modern civilization as we know it since modern societies have become incredibly dependent on the functioning of key infrastructure to survive.

War as a state practice has been delegitimized since the end of the First World War with the war guilt clause in the Treaty of Versailles, which declared that Germany was solely responsible for the war.[2] After the end of the Second World War, waging a war of aggression became a war crime in itself, practically making it illegal for any government to use force internationally in other than narrowly defined circumstances (as self-defense, under the authority of the UN Security Council, and when another government requested security assistance). The practice of interstate war has been in decline since 1945, and the wars that still take place tend to be far less violent than the wars of the pre-1945 period if adjusted for overall population size.[3]

Michael Mandelbaum argued that '[m]ajor war is obsolete in the way that slavery, dueling or foot-binding are obsolete: it is a social practice that was once considered normal, useful—and even desirable—but that now seems odious.'[4] Major conventional war has not completely disappeared, as indicated in Russia's invasion of Ukraine from February 2022, but it remains doubtful that the world can return to the kinds of major war seen before 1945. Does this mean that the world is now headed for a period of Kantian eternal peace and a world free from war and violence, as proclaimed by those who believe in a civilizing process?[5] Hardly. The argument presented in this book is that war has evolved into something that is not easily recognizable as war, but that is no less decisive or consequential.

Clausewitzian War

The Western understanding of war has been greatly shaped by the ideas of Carl von Clausewitz.[6] According to Clausewitz, war has three defining features: War

DOI: 10.4324/9781003396963-1

must be violent ('an act of force to compel the enemy to do our will'); the violence in war is instrumental or a means to an end (war contains 'an element of subordination, as an instrument of policy, which makes it subject to reason alone'); and war must be political in nature ('war is a mere continuation of politics by other means').[7] In the Clausewitzian paradigm, there cannot be such a thing as a nonviolent war or a non-political war. Furthermore, the Western view of war is based on a strict dichotomy of war and peace in which peace is defined as the suspension of the use of violence.[8] This means that war is easily recognizable as such through the instrumental and overt use of organized violence for political objectives by identified belligerents, with a clearly identifiable start point and end point, typically marked by a ceasefire and a subsequent peace agreement.

Clausewitz suggested that war is a contest of wills and 'nothing but a duel on a larger scale' fought by armed forces.[9] This means that, in the Western understanding of war, wars are fought and decided by military forces, whether regular or irregular. This makes the destruction of these military forces on the battlefield the key, if not the only, path to victory, since diminishing the ability of the enemy to fight would weaken the enemy's will. Western militaries have become exceedingly proficient in their capability to destroy enemy forces on the battlefield owing to superior technology such intelligence, surveillance, and reconnaissance (ISR) and precision-guided munitions. Not surprisingly, Western governments and militaries cling to a traditional view of war, even though they are confronted with unconventional challenges and with novel approaches to conducting conflict.

War without Rules

It has been noted that the West's main adversaries have a somewhat different view of war that deliberately blurs established distinctions of war and peace.[10] They see themselves in a constant state of war and conflict with the West in which military action may occur, but which is primarily fought by non-military means, such as informational, economic, and covert means. Their purpose is to systematically weaken the adversary, which can, over time, provide them with a decisive advantage that either enables a military victory at a late stage of the conflict or makes the use of military force unnecessary owing to political and economic subjugation.

The appearance of conflict today is complex, if not outright confusing. Instead of direct military threats, Western powers are now threatened by destabilizing propaganda and disinformation, by the disruption of critical supply chains through economic warfare, by cyberattacks against critical infrastructure, by terrorism and sabotage, by influxes of large numbers of migrants, by cultural subversion, by gray zone tactics such as the abuse of international law, by mysterious pandemics, and by their societies crumbling through widespread drug addiction and lawlessness. Some of this has been described as unrestricted warfare, hybrid warfare, gray zone conflict, stealth war, twilight struggle, the weaponization of everything, or cognitive warfare.[11] This book suggests that the theory of fifth generation warfare (5GW) may provide the best theoretical framework, up to now, for understanding the new paradigm of war, which, interestingly, coexists with the older paradigm.

What Is War and Can There Be Nonviolent or Non-Military War?

Traditionalists reject the idea that war can be anything other than an armed conflict and believe that the focus of strategy needs to be on the use of force and not on non-military initiatives and nonviolent threats. For example, Donald Stoker and Craig Whiteside have argued that concepts such as gray zone conflict and hybrid war confuse the distinction between war and peace and undermine U.S. strategic thinking with sloppy theories that do not hold up historically and against existing theory. They define war as 'a distinct state in which violence is used to achieve political ends.'[12] They claim that

> by confusing competition among adversaries with things called *hybrid* or *gray-zone war*, we risk conflating everything with war—a dangerous proposition. If we are at war with another country, our citizens rightly can ask what exactly we are doing about it. If it is merely heated competition and international politics, meaning who gets what, when, and where, then elements of national power other than force or threats to use force will have to be relied on to a larger degree—and this seems to be the root of American leaders' problem.[13]

In a legal and technical sense and from a policy perspective, Stoker and Whiteside are correct. Under international law, war is defined through a number of related concepts such as force or act of force (including the threat of force), aggression, armed attack, and armed conflict. The legal consequence of an act of force, aggression, or armed attack is that a state that has suffered such can invoke the right to self-defense as guaranteed under Article 51 of the UN Charter and respond with proportionate force against the aggressor.[14] International law does not allow the use of force for any actions that fall short of being an act of force, which includes pretty much all the non-military means used by adversaries in conflicts short of war, such as espionage, subversion, economic sanctions, lawfare, engineered mass migration, and so on.

Stoker and Whiteside are also correct about the intellectual weaknesses of some current theories of war and conflict. There is also a concerning tendency by politicians and the media to invoke the war metaphor in relation to all kinds of social issues and challenges. Mark Galeotti pointed out that '[s]uddenly, everything can be weaponised as part of the expanding array (arsenal, even) of military metaphors all around us' and that 'every war since one gang of cavemen squared off against another over possession of the driest caves has been "hybrid."'[15] In short, Galeotti suggests that belligerents have always used all means at their disposal to harm their adversaries, and that this has never been a big deal.

Where the critics of a more expansive concept of war are wrong is the failure to consider war not only in terms of the means used (force or violence) but also in terms of the outcomes (political coercion and subjugation). Of course, it would be wrong to characterize any kind of peaceful competition as a potential act of war or as a war analog, as this does make the concept of war meaningless. At the same time, it would be equally wrong and foolish to disregard new modes of attack that are covert and potentially less or not violent as inconsequential or far less than war, if they can plausibly destroy a nation from within and result in the de facto subjugation of a population by external actors cooperating with domestic actors. While this approach of exploiting subversion, elite treachery, and corruption is by no means a novelty in military history, it is also the case that modern societies are perhaps far more vulnerable to indirect modes of attack than societies were in previous historical periods.

Is 5GW Just a Conspiracy Theory?

Besides the criticism that nonviolent war is not war, there is also the concern that 5GW is little more than a conspiracy theory since it emphasizes that many acts of aggression may be secretive, rely on the covert manipulation of proxies, and be conducted at a societal level. The essence of 5GW is covert psychological warfare, and obviously the perpetrators of the usually subtle psychological attacks have deniability: It is generally difficult to prove that psychological manipulation has occurred or to convince victims of such that they have been manipulated. Like brainwashed cult members, they will stubbornly insist that all their actions were voluntary and reasonable.[16] As a result, there is a tendency to dismiss the notions of 5GW and psychological/cognitive warfare targeting entire societies as conspiracy theorizing.

It shall be noted that the term conspiracy theory has come to be 'a pejorative label for ideas that other people think are outlandish.'[17] As Lance deHaven-Smith pointed out, conspiracy theory was also a term invented by the CIA's psychological warfare experts to deflect from the inconsistencies in the government's claims about the Kennedy assassination.[18] What matters here is that political conspiracies are far from impossible and have occurred with some regularity in history.[19] Espionage and covert action are inherently conspiratorial activities that involve secret meetings, secret nefarious plots, and hidden actors. Nobody can deny that both are features of modern statecraft. Where opinions differ is simply as to how effective or threatening these activities are to targeted societies. According to research by Lindsay O'Rourke, the U.S. government succeeded in overthrowing/changing at least a dozen governments in the Cold War period using covert means.[20] Arguably, the destabilization of the Soviet Union by the policies (overt and covert) of the Reagan administration contributed greatly to the superpower's collapse in the late 1980s.[21] It would be naïve to believe that the United States is not equally vulnerable to similar methods of destabilization used by others.

Furthermore, Western strategic thinking seems to be lagging behind the strategic thinking of the Russian and Chinese state security establishments. Russian theorists have adopted the term 'information warfare' to describe methods of nonviolent destabilization. Russian theorists consider information a dangerous weapon. According to Jolanta Darczewska, most *Russian authors understand 'information warfare' as influencing the consciousness of the masses as part of the rivalry between the different civilisational systems adopted by different countries in the information space by use of special means to control information resources as 'information weapons'*. They thus mix the military and non-military order and the technological (*cyberspace*) and social order (*information space*) by definition, and make direct references to 'Cold War' and 'psychological warfare' between the East and the West.[22]

NATO analysts have become increasingly concerned about Russian attempts to create divisions between Western societies in order to destabilize governments. NATO analysts fear that '[s]everal successive campaigns could be launched with the long-term objective of disrupting entire societies or alliances, by seeding doubts about governance, subverting democratic processes, triggering civil disturbances, or instigating separatist movements.'[23] If things get progressively worse in a country, with economic decline, increasing internal polarization, massive migrant influxes, failing critical infrastructure, and worsening lawlessness in major cities, one could chalk it up to bad luck or one could try to figure out what might be happening by seriously considering the possibility that some of the bad things may be deliberate, may be secretly orchestrated, and may be part of an overall strategy by some external actor, perhaps working with domestic groups and traitors. Fifth generation warfare can provide a useful framework for such analysis.

Fifth Generation Warfare

Fifth generation warfare is a term that was apparently first coined by Robert David Steele in 2003.[24] The primary idea is that 5GW represents an evolution of the older theory of fourth generation warfare (4GW), which was introduced in an influential paper by William Lind et al., originally published in the *Marine Corps Gazette* in 1989.[25] The proponents of 4GW argued that warfare had undergone successive generations from 1800, with the current, fourth generation being focused on political warfare and evolved insurgencies.[26] In contrast, the emerging fifth generation would be focused on influence and psychological warfare, with a much diminished role of violence. Fifth generation warfare skips the battlefield and directly targets the main objective of war—'to compel the enemy to do our will.' After all, 'to subdue the enemy without fighting is the acme of skill' in the view of Sun Tzu.[27] According to Shane Deichman, '[t]he ability to shape the perception—and therefore the opinions—of a target audience is far more important than the ability to deliver kinetic energy, and will determine the ultimate victor in tomorrow's wars.'[28]

Fifth generation warfare is very much like a confidence trick as portrayed in the Paul Newman and Robert Redford movie *The Sting* (1973). The con artists create an alternate reality in which they trick a mobster into betting all of his money on a rigged horse race. As the con unfolds, the mobster fails to understand that he was defrauded, and he is even happy to get away from the scene when he seemingly lost total control over the situation. As Daniel Abbott argued, '[a] fifth Generation War might be fought with one side not knowing who it is fighting. Or

even, a brilliantly executed 5GW might involve one side being completely igno-
rant that there ever was a war.'[29] Because of the great reliance on deception and
psychological manipulation of adversaries to achieve victory, 5GW is the most
secretive mode of warfare that exists, which makes it extremely difficult to study
and to understand.[30]

Several definitions of 5GW have been proposed by different authors, some of
which are contained in Daniel Abbott's *The Handbook of 5GW*.

> 5GW is the manipulation of observational context in order to make the en-
> emy do our will. Since an act of force is not required to manipulate observa-
> tional context, force is not required to wage 5GW. Since 5GW is undoubtedly
> war, war must be more than a mere act of force.[31]

Abbott proposed the following definition of 5GW:

> An emerging theory of warfare premised upon manipulation of multiple eco-
> nomic, political, social and military forces in multiple domains to effect po-
> sitional changes in systems and achieve consilience of effects to leverage a
> specific goal or set of circumstances.[32]

Approached from different angles and as suggested by different 5GW proponents, it can be stated that 5GW has the following defining characteristics:

- Fifth generation warfare focuses on the manipulation of perception, identity,
  and context, and its primary means are the tools of psychological warfare and
  social engineering.
- Fifth generation warfare bypasses the battlefield and targets society as a whole,
  rather than its military forces.
- Violence in 5GW is very dispersed or hidden, which makes it difficult to
  perceive as war.
- Fifth generation warfare relies on covert or ambiguous means that hide nefarious activities or disguise them as benign or harmless.
- Fifth generation warfare can be waged by super-empowered individuals and
  small groups against societies and states, often by co-opting states or other proxies, using state resources, and having state-like capabilities at their disposal.
- The objective of 5GW is the destruction of the existing political and societal
  order with the purpose of bringing it into alignment with the ideological or religious objectives of the aggressor.
- Fifth generation warfare may degrade to 4GW in the event a state or society
  collapses, or it may evolve to a higher level such as sixth generation warfare
  (6GW).
- Fifth generation warfare is enabled by key technologies that attack and steer
  societies, including information technology, and also, increasingly, emerging
  technologies such as biotechnology, neurotechnology, artificial intelligence
  (AI), and nanotechnology.

The author, hence, proposes the following definition of 5GW, which will be used in this book:

> 5GW is a distinctive understanding of conflict that occurs at the societal level. It is driven by sub-nation state actors (rather than states or governments) against a government or a society for ideological or nonpolitical reasons, using covert means, such as psychological manipulation, and ambiguous means, such as cultural influence, rather than overt kinetic means. The objective of 5GW is to overthrow an existing political order or change the culture in a society in accordance with the goals of the aggressor. 5GW is not replacing older or more established modes of conflict, but it clearly shifts the emphasis from the control of physical terrain to the control of the human terrain and the human mind as a target of attack. The objective in 5GW is ultimately influence over, if not control of, the human mind and human behavior on a larger scale. 5GW can degenerate into 4GW or 3GW, at which point kinetic action will be more decisive.

The theory of 5GW as proposed by Abbott et al. is not incompatible with the enduring existence of older generations of warfare alongside 5GW. In other words, the fact that a major conventional war is taking place in Ukraine is not evidence that 5GW would be now irrelevant or that the nature of war has remained unchanged. As will be shown in later chapters, kinetic action, or the ability to destroy forces on the battlefield, is far less important than the ability to resist and counter non-military and nonviolent modes of attack. As of now, the ultimate objective of war on a societal scale is the domination of the human domain and the ability of a belligerent to undermine the adversary's will to resist or to engage in military conflict. This approach goes beyond the political warfare of the Cold War period, which already foreshadowed 5GW in many ways.[33]

Chapter Overview

The first chapter will introduce the concept of generational warfare and explain the 4GW school of thought as it emerged in the 1990s and early 2000s. The second chapter will explore some of the key concepts and theories developed in the 1990s that are the foundations of the 5GW theory. The third chapter will connect 5GW to the counterinsurgency paradigm, which is about shaping the operational environment to one's advantage in relation to the human terrain. The fourth chapter will discuss cultural and cognitive warfare approaches and methods from within the 5GW framework by discussing four distinct examples of 5GW. The fifth chapter will explore the role of emerging technologies in 5GW, which could make 5GW particularly dangerous. The final chapter will discuss the future of warfare and some ideas that could define sixth generation warfare, namely attacks against infrastructure and other technological systems as a new type of siege warfare that collapses a society from within by attacking technical systems.

Notes

1 Helen Caldicott, *The New Nuclear Danger: George W. Bush's Military–Industrial Complex* (New York: The New Press, 2002), 7–12.
2 Leslie C. Green, *The Contemporary Law of Armed Conflict* (Manchester: Manchester University Press, 2000), 4.
3 Steven Pinker, *The Better Angels of Our Nature: Why Violence Has Declined* (New York: Penguin Books, 2011), 230 and 304.
4 Michael Mandelbaum, 'Is Major War Obsolete?' *Survival* 40, no. 4 (1998): 34.
5 Steven Pinker argued that the world has seen an overall and largely consistent decline in violence over the last few hundred years, which would suggest that the future would be more peaceful. Pinker, *The Better Angels of Our Nature*.
6 Martin van Creveld, *The Transformation of War* (New York: The Free Press, 1991), 34.
7 Thomas Rid, *Cyber War Will Not Take Place* (Oxford: Oxford University Press, 2013), 1–2.
8 Félix E. Martín, 'Critical Analysis of the Concept of Peace in International Relations,' *Peace Research* 37, no. 2 (2005): 45.
9 Carl von Clausewitz, *On War: Translated and Edited by Michael Howard and Peter Paret* (Princeton, NJ: Princeton University Press, 1976), 75.
10 Daniel Burkhardt and Alison Woody, 'Strategic Competition beyond Peace and War,' *Joint Forces Quarterly* 86, no. 3 (2017): 20–27.
11 Liang Qiao and Xiangsui Wang, *Unrestricted Warfare* (Naples, Italy: Albatross, 2020); Frank G. Hoffman, *Conflict in the 21st Century: The Rise of Hybrid Wars* (Arlington, VA: Potomac Institute, 2007); Michael J. Mazarr, *Mastering the Gray Zone: Understanding a Changing Era of Conflict* (Carlisle Barracks, PA: Army War College Press, 2015); Robert Spalding, *Stealth War: How China Took Over while America's Elite Slept* (New York: Portfolio/Penguin, 2019); Hal Brands, *The Twilight Struggle: What the Cold War Teaches Us about Great-Power Rivalry Today* (New Haven, CT: Yale University Press, 2022); Mark Galeotti, *The Weaponisation of Everything: A Field Guide to the New Way of War* (New Haven, CT: Yale University Press, 2022); and B. Claverie Brébot, N. Buchler, and F. Du Cluzel, 'Cognitive Warfare: First NATO Scientific Meeting on Cognitive Warfare,' Symposium in Bordeaux, France, June 21, 2021.
12 Donald Stoker and Craig Whiteside, 'Gray Zone Conflict and Hybrid War—Two Failures of American Strategic Thinking,' *Naval War College Review* 73, no. 1 (2020): 29.
13 Stoker and Whiteside, 'Gray Zone Conflict and Hybrid War,' 44; original italics.
14 Green, *The Contemporary Law of Armed Conflict*, 9.
15 Galeotti, *The Weaponisation of Everything*, 11.
16 Steven Hassan, *Combating Cult Mind Control* (Newton, MA: Freedom of Mind Press, 2018), 113.
17 G.B. Arnold, *Conspiracy Theory in Film, Television, and Politics* (Westport, CT: Praeger, 2008), 20, quoted from Lance DeHaven-Smith, *Conspiracy Theory in America* (Austin, TX: University of Texas Press), 40.
18 DeHaven-Smith, *Conspiracy Theory in America*, 21.
19 Even proponents of the conspiracy theory label such as Cass Sunstein have to acknowledge that 'some conspiracy theories have turned out to be true,' such as MK ULTRA, Operation Northwoods, and Watergate. Cass Sunstein and Adrian Vermeule, 'Conspiracy Theories: Causes and Cures,' *The Journal of Political Philosophy* 17, no. 2 (2009): 206.
20 Lindsay O'Rourke, *Covert Regime Change: America's Secret Cold War* (Ithaca, NY: Cornell University Press, 2018), 109.
21 Peter Schweizer, *Victory! The Reagan Administration's Secret Strategy That Hastened the Collapse of the Soviet Union* (New York: The Atlantic Monthly Press, 1994), XII–XIX.

22 Jolanta Darczewska, 'The Anatomy of Russian Information Warfare: The Crimean Operation, a Case Study,' *OSW: Point of View* 42 (May 2014): 12; original italics.
23 NATO, 'Countering Cognitive Warfare: Awareness and Resilience,' NATO website, May 20, 2021, available at: www.nato.int/docu/review/articles/2021/05/20/countering-cognitive-warfare-awareness-and-resilience/index.html
24 Daniel H. Abbott (ed.), *The Handbook of 5GW* (Ann Arbor, MI: Nimble Books, 2010), 220.
25 William S. Lind, Keith Nightingale, John F. Schmitt, Joseph W. Sutton, Gary I. Wilson, 'The Changing Face of War: Into the Fourth Generation,' *Marine Corps Gazette* (March 2016): 86–90.
26 Thomas X. Hammes, *The Sling and the Stone: On War in the 21st Century* (Paul, MN: Zenith Press, 2004).
27 Sun Tzu, *The Art of War: Translated by Ralph D. Sawyer* (New York: Basic Books, 1994), 175–180 (Chapter 3).
28 Shane Deichman, 'Battling for Perception: Into the 5th Generation?' in: Abbott, *The Handbook of 5GW*, 19.
29 Daniel H. Abbott, 'Go Deep: OODA and the Rainbow of xGW,' in: Abbott, *The Handbook of 5GW*, 180.
30 Daniel H. Abbott, 'Introduction,' in Abbott, *The Handbook of 5GW*, 10.
31 L.C. Rees, 'The End of the Rainbow: Implications of 5GW for a General Theory of War,' in: Abbott, *The Handbook of 5GW*, 21.
32 Daniel H. Abbott, 'The Terminology of xGW,' in: Daniel H. Abbott, *The Handbook of 5GW*, 206.
33 William R. Kintner and Joseph Z. Kornfeder, *The New Frontier of War: Political Warfare, Present and Future* (Chicago, IL: Henry Regnery, 1962).

References

Abbott, Daniel H. (ed.), *The Handbook of 5GW* (Ann Arbor, MI: Nimble Books, 2010).
———, 'Introduction,' in: Daniel H. Abbott, *The Handbook of 5GW* (Ann Arbor, MI: Nimble Books, 2010), 6.
———, 'Go Deep: OODA and the Rainbow of xGW,' in: Daniel H. Abbott, *The Handbook of 5GW* (Ann Arbor, MI: Nimble Books, 2010), 174–186.
———, 'The Terminology of xGW,' in: Daniel H. Abbott, *The Handbook of 5GW* (Ann Arbor, MI: Nimble Books, 2010), 204.
Brands, Hal, *The Twilight Struggle: What the Cold War Teaches Us about Great-Power Rivalry Today* (New Haven, CT: Yale University Press, 2022).
Brébot, B. Claverie, Buchler, N., and Du Cluzel, F., 'Cognitive Warfare: First NATO Scientific Meeting on Cognitive Warfare,' Symposium in Bordeaux, France, June 21, 2021.
Burkhardt, Daniel and Woody, Alison, 'Strategic Competition beyond Peace and War,' *Joint Forces Quarterly* 86, no. 3 (2017): 20–27.
Caldicott, Helen, *The New Nuclear Danger: George W. Bush's Military–Industrial Complex* (New York: The New Press, 2002).
Darczewska, Jolanta, 'The Anatomy of Russian Information Warfare: The Crimean Operation, a Case Study,' *OSW: Point of View* 42 (May 2014).
DeHaven-Smith, Lance, *Conspiracy Theory in America* (Austin, TX: University of Texas Press).
Deichman, Shane, 'Battling for Perception: Into the 5th Generation?' in: Daniel H. Abbott, *The Handbook of 5GW* (Ann Arbor, MI: Nimble Books, 2010), 11–19.
Galeotti, Mark, *The Weaponisation of Everything: A Field Guide to the New Way of War* (New Haven, CT: Yale University Press, 2022).

Green, Leslie C., *The Contemporary Law of Armed Conflict* (Manchester: Manchester University Press, 2000).

Hammes, Thomas X., *The Sling and the Stone: On War in the 21st Century* (Paul, MN: Zenith Press, 2004).

Hassan, Steven, *Combating Cult Mind Control* (Newton, MA: Freedom of Mind Press, 2018).

Hoffman, Frank G., *Conflict in the 21st Century: The Rise of Hybrid Wars* (Arlington, VA: Potomac Institute, 2007).

Kintner, William R. and Kornfeder, Joseph Z., *The New Frontier of War: Political Warfare, Present and Future* (Chicago, IL: Henry Regnery, 1962).

Lind, William S., Nightingale, Keith, Schmitt, John F., Sutton, Joseph W., and Wilson, Gary I., 'The Changing Face of War: Into the Fourth Generation,' *Marine Corps Gazette* (March 2016): 86–90.

Mandelbaum, Michael, 'Is Major War Obsolete?' *Survival* 40, no. 4 (1998): 20–38.

Martín, Félix E., 'Critical Analysis of the Concept of Peace in International Relations,' *Peace Research* 37, no. 2 (2005): 45–59.

Mazzar, Michael J., *Mastering the Gray Zone: Understanding a Changing Era of Conflict* (Carlisle Barracks, PA: Army War College Press, 2015).

NATO, 'Countering Cognitive Warfare: Awareness and Resilience,' NATO website, May 20, 2021, available at: www.nato.int/docu/review/articles/2021/05/20/countering-cognitive-warfare-awareness-and-resilience/index.html

O'Rourke, Lindsay, *Covert Regime Change: America's Secret Cold War* (Ithaca, NY: Cornell University Press, 2018).

Pinker, Steven, *The Better Angels of Our Nature: Why Violence Has Declined* (New York: Penguin Books, 2011).

Qiao, Liang and Wang, Xiangsui, *Unrestricted Warfare* (Naples, Italy: Albatross, 2020).

Rees, L.C., 'The End of the Rainbow: Implications of 5GW for a General Theory of War,' in: Daniel H. Abbott, *The Handbook of 5GW* (Ann Arbor, MI: Nimble Books, 2010), 20–29.

Rid, Thomas, *Cyber War Will Not Take Place* (Oxford: Oxford University Press, 2013).

Schweizer, Peter, *Victory! The Reagan Administration's Secret Strategy That Hastened the Collapse of the Soviet Union* (New York: The Atlantic Monthly Press, 1994).

Spalding, Robert, *Stealth War: How China Took Over while America's Elite Slept* (New York: Portfolio/Penguin, 2019).

Stoker, Donald and Whiteside, Craig, 'Gray Zone Conflict and Hybrid War—Two Failures of American Strategic Thinking,' *Naval War College Review* 73, no. 1 (2020): 19–54.

Sunstein, Cass and Vermeule, Adrian, 'Conspiracy Theories: Causes and Cures,' *The Journal of Political Philosophy* 17, no. 2 (2009): 202–227.

Sun Tzu, *The Art of War: Translated by Ralph D. Sawyer* (New York: Basic Books, 1994).

Van Creveld, Martin, *The Transformation of War* (New York: The Free Press, 1991).

Von Clausewitz, Carl, *On War: Translated and Edited by Michael Howard and Peter Paret* (Princeton, NJ: Princeton University Press, 1976).

1 Fourth Generation Warfare

This chapter discusses the concept of fourth generation warfare as it was developed by William Lind, Martin van Creveld, and Thomas Hammes.[1] They represent three major reformulations of 4GW with Lind et al. first proposing the concept in the late 1980s, van Creveld adding the idea of 'nontrinitarian war' as a likely future of war in the early 1990s, and Hammes using 4GW as a framework for his analysis of contemporary insurgencies in the early 2000s.[2] Although van Creveld does not explicitly refer to 4GW in his writing, he is generally considered to be one of the leading thinkers in this school of thought.[3] Hammes based his theory of insurgency on the original ideas of Lind et al. and adapted it to the particular strategic challenges after 9/11. In short, the 4GW school proclaims that state militaries are operating within a paradigm of war that was shaped in the 19th and 20th centuries and that this would be ill suited for dealing with a new generation of war that aims to exploit the limitations of regular armed forces or asymmetry.

During the 1990s, this unconventional challenge was also referred to as 'asymmetric warfare' in the context of the revolution in military affairs (RMA) school of thought.[4] Where the RMA school and the 4GW school disagreed was on the ability of regular armed forces to successfully fight and defeat irregular adversaries. Whereas the proponents of an RMA envisioned that technology—in particular, advanced intelligence, surveillance, and reconnaissance (ISR)—could provide conventional forces with a decisive advantage over their irregular adversaries, the 4GW school claimed that advanced technology offered no significant advantage, and that only more radical changes to the organization and doctrine of modern armed forces could provide the best chances for defeating 4GW insurgencies. In short, the 4GW school was motivated by understanding new features of insurgencies that made them, in its view, so much more difficult for governments to suppress.

The Decay of the State

What became a key argument of the proponents of 4GW in the 1990s was the supposed decline of the state and the Westphalian international system based on sovereign nation states. The Israeli military historian Martin van Creveld argued, in *The Transformation of War* and later in *The Rise and Decline of the State*, that the sovereign nation state was a fairly recent invention and might not survive in

DOI: 10.4324/9781003396963-2

the long term.[5] Van Creveld's reasoning was the following: In the Middle Ages, the political order was very complex owing to the existence of a large number of princedoms, city states, and kingdoms, with overlapping claims of authority made by empires and the Catholic Church. Rulers treated the territories that they controlled like private possessions, and they were competing with private actors such as guilds, corporations, mercenary armies, pirates, and bandits. The medieval political order came to an end with the establishment of the Westphalian system based on the idea of the sovereignty of nation states. Nation states were able to centralize power and create large bureaucracies that could raise and collect taxes much more effectively than was possible under the feudal system.[6]

The consequence was that states now could afford to maintain large standing armies and could easily outcompete nonstate actors in the arena of warfare. In particular, states could afford, among other things, the construction of more advanced fortifications that could withstand attacks with cannons.[7] The result was that nonstate actors were gradually suppressed by more powerful states, which allowed the states to gradually establish a monopoly of force in their territory.[8] The successful suppression of nonstate violence hence enabled the modern nation state as a sovereign entity within an international system of states.

Trinitarian War

The rise of the sovereign nation state has led to what van Creveld calls 'trinitarian war.' Trinitarian war is essentially the Clausewitzian understanding of war as war among states for the purpose of achieving a limited political objective. The Clausewitzian trinity would consist of a government that determines the political objectives, regular armed forces that fight the war, and the people, who indirectly support the war through participation in the economy.[9] It was up to the sovereign whether or when to go to war, and only the representatives of the sovereign could lawfully engage in combat.

As state militaries became professionalized in the 19th century, it meant that they were more carefully recruited, better trained, and more tightly controlled so that they would not engage in unauthorized actions that could embarrass the sovereign. Soldiers 'were supposed to fight only while in uniform, carrying their arms "openly," and obeying a commander who could be held accountable for their actions.'[10] Only soldiers acting on behalf of a sovereign state could lawfully participate in hostilities, while civilians were not permitted to do so. In other words, war became, from the 19th century to the middle of the 20th century, the almost exclusive domain of nation states.

The conduct of war has become much more regulated through international law, which has defined permissible practices in war and sought to protect civilian populations, who were to be left unharmed. These rules of war were codified in the Lieber Code of 1863, the Hague Conventions of 1899 and 1907, and the Geneva Conventions of 1864, 1929, and 1949. Trinitarian war culminated in the total wars of the 20th century, the First World War and the Second World War.

Since 1945, there has been a notable decline in interstate war owing to two major developments—one technological and one social. Nuclear weapons have made great power war inconceivable because of the tremendous destruction that would follow, which effectively undermines the very idea of a limited political objective of war.[11] Second, as Francis Fukuyama pointed out, interstate war has been delegitimized as a state practice and it no longer offers much of an advantage. Since 'resources can be obtained peacefully through a global system of free trade, war makes much less economic sense than it did two or three hundred years ago.'[12] The wars that still take place are predominantly intrastate wars, which are caused mostly by the internal weakness of states in parts of the world where nation states are far less established and are largely artificial constructs created by former colonial powers.[13] Intrastate conflicts featuring insurgencies, terrorism, and lawlessness and broadly characterized as 'low-intensity conflicts' have become the dominant form of war, with three-quarters of the 160 or so armed conflicts from 1945 to 1990 falling into this category.[14]

Guerrilla Warfare

An early challenge to the concept of Clausewitzian war was the emergence of the modern guerrilla in the period of the Napoleonic Wars. Carl Schmitt pointed out that '[t]he partisan of the Spanish guerrilla war of 1808 was the first who dared to fight irregularly against the first modern regular army.'[15] What was new about the irregular resistance to Napoleon's army was the aspect of nationalism that motivated the guerrillas, as no legitimate authority, such as the Spanish government or nobility, was available to direct them. At the peak of the war, the French committed over 400,000 troops to defeating the Spanish guerrillas but were unable to do so because of British support given to the guerrillas and because of their ability to harass supply lines and attack isolated garrisons.[16] The French suffered substantial losses in Spain, and Napoleon was forced to retreat from Spain in 1813.

In the 20th century, guerrilla warfare characterized the struggle against colonialism in developing countries.[17] Liberation and revolutionary movements were able to defeat colonial powers and their colonial governments in Indochina, Israel, Kenya, Algeria, Vietnam, Mozambique, among others. Guerrilla forces are by definition militarily too weak to defeat their regular adversaries in decisive battles and, therefore, must focus on the political dimension of the conflict while avoiding a military defeat. Henry Kissinger famously argued, in 1969, that 'the guerrilla wins if he does not lose. The conventional army loses if it does not win.'[18] As a result, government forces must seek to suppress an insurgency quickly in order to win, while guerrillas need to drag out the conflict as long as they can. Guerrilla warfare is, therefore, by its very nature protracted war, since guerrilla forces need to either wear down the will of an occupying power by increasing the political and economic cost to it or to slowly build up their strength over time to be eventually strong enough to take on the government forces directly in a conventional battle.[19]

A key characteristic of guerrilla warfare is that the guerrilla fights irregularly, meaning the guerrilla violates conventions of war by not acting under the control of a state, by not wearing a uniform, and by engaging in combat actions that amount to terrorism from the perspective of their conventional adversaries. For a government threatened by an insurgency, the guerrilla is nothing but a criminal engaged in terrorism.[20] Although guerrillas were later given a legal status in the Hague Convention of 1907, it has done little to change the nature of irregular conflict and how it is perceived by a government. In short, the guerrilla fights unfairly as '[t]he partisan is still the one who refuses to carry weapons openly, who fights from ambush, and who uses the enemy's uniform, as well as true or false insignias and every type of civilian clothing as camouflage.'[21] Guerrillas have nothing to gain and much to lose from trying to comply with the laws of armed conflict: It is unlikely to result in better treatment by their opponents when captured and it severely diminishes their advantages.

Neomedievalism

Sean McFate argued that '[t]he world order may be slowly returning to the status quo ante of the Middle Ages.'[22] Neomedievalism, as 'a non-state-centric, multipolar international system of overlapping authorities and allegiances within the same territory,' could possibly replace the established Westphalian order.[23] Intuitively, this seems like a plausible assumption. The post-Cold War world order has clearly become more chaotic with the power of major states declining, the rise of various nonstate actors, both legitimate and illegitimate, and the proliferation of civil wars and political instability across the globe. Philip Cerny observed that

> [a] kind of generalised 'insecurity from below' has emerged, bound up with the dual character of globalisation as a global–local dialectic, whipsawing the state between the international and transnational, on the one hand, and an increasingly complex set of micro- and meso-level phenomena on the other—what Rosenau has called 'fragmegration.'[24]

The states are, hence, trapped in a dynamic that erodes their sovereignty from both below and above.

McFate invoked Hedley Bull's concept of neomedievalism developed in his book *The Anarchical Society* of 1977. Bull had identified five necessary conditions for a return to medievalism: Technological unification, regional integration, rise of transnational organizations, disintegration of states, and the return of private international violence.[25] McFate argued that these conditions are largely met by a number of trends, most importantly globalization, which has spearheaded a technological unification of the world. The world would now be much more connected economically, financially, informationally, and culturally. The internet in particular makes it possible to penetrate other societies from afar with ease, to spread information. National borders have become much harder to police owing to the free flow of capital, goods, and people. States have become more regionally integrated because of the creation of international organizations that now compete with state sovereignty. McFate mentioned in particular the International Criminal Court (ICC), which can investigate and prosecute war criminals and even heads of state responsible for government policies that violate the UN Charter.[26]

Since 1945, there has also been a massive growth in NGOs that operate internationally, with no or little government control. Most of the NGOs are set up as charitable organizations or single-issue groups that seek to influence government policies internationally. Notable are organizations such as Amnesty International, Greenpeace, Human Rights Watch, or CARITAS. They are sometimes funded or hired by governments to respond to humanitarian catastrophes or to assist with development and democratization. Global multinational organizations now rival governments in terms of the financial resources under their control and they have managed to exploit national borders to their advantage by shifting production to countries that offer cheaper labor and better tax conditions.[27] Even illegitimate nonstate actors such as contemporary terror groups, insurgents, and criminal organizations have corroded state sovereignty and they no longer see any benefit in establishing or controlling a state themselves.[28]

A consequence of these trends is the disintegration of weak states, which are increasingly unable to contain or control violent nonstate actors. War has become privatized, and with it there are now economic rather than political motivations for conducting war. This has led to a new globalized war economy, described by Mary Kaldor, which relies largely on external sources for financing war such as extorted ransoms, control of food and other necessities, war taxes imposed by insurgents, the illegal trade of drugs and other commodities, assistance from other governments, exploitation of foreign aid, and support from diasporas.[29] Wars are endlessly prolonged, which creates failing states and failed states. Robert Kaplan, therefore, predicted a coming anarchy where much of modern civilization collapses and Western cities come to resemble developing countries, with extreme poverty and great wealth existing side by side.[30] He wrote,

> West Africa is becoming the symbol of worldwide demographic, environmental, and societal stress, in which criminal anarchy emerges as the real 'strategic' danger. Disease, overpopulation, unprovoked crime, scarcity of resources, refugee migrations, the increasing erosion of nation-states and international borders, and the empowerment of private armies, security firms, and international drug cartels are now most tellingly demonstrated through a West African prism.[31]

Ralph Peters observed the emergence of a 'new warrior class' that was unconstrained by Western morality and, hence, created dilemmas as to how Western militaries could fight it.[32] Steven Metz and James Kievit went so far as to claim that

> [i]n the post-Cold War era, many if not most Third World states will fragment into smaller units. Ungovernability and instability will be the norm. Even those which formally remain intact will see political and military power

dispersed among warlords, primal militias, and well-organized politico-criminal organizations.[33]

In other words, wars would no longer be primarily fought by states.

Private Armies

Mercenaries have always been a feature of war, although they were hardly significant in the major wars of the 19th and 20th centuries, with the exception of some colonial wars in Congo (1960–1965), the Biafra War (1967–1970), and the Angolan Civil War (1975–1976). The participation of white mercenaries in African wars was so notorious in the 1960s and 1970s that the Organization of the African Union passed a convention that declared 'mercenarism' a crime.[34] Mercenary companies were typically shadowy outfits established for a particular war and closed down after the war ended. By the early 1990s, the character of the mercenary firms changed as they now intended to permanently stay in business and to achieve reputability by marketing themselves as military service providers to legitimate governments facing internal or external threats.[35]

Peter Singer observed that there were both supply-side and demand-side factors that helped the rapid growth of private military companies (or private military firms).[36] On the one hand, militaries around the world were downsizing after the end of the Cold War, which created a surplus of skilled military professionals looking for employment, as well as a surplus of modern military equipment (mostly from the former Soviet Union) that became available to the highest bidder. On the other hand, there were also new conflicts breaking out or intensifying that were previously contained by the superpowers, most notably in the Balkans, Angola, Sierra Leone, Congo, Ethiopia–Eritrea, and others. This was combined with the weakness of states and a lack of an international response to violent nonstate actors.[37]

The South African company Executive Outcomes (EO) was among the first to usher in a new privatized military industry in the 1990s. EO intervened in two civil wars on behalf of the respective governments. In the early 1990s, the Angolan MPLA government was facing a severe threat from the rebel group UNITA, led by Jonas Savimbi, which led to a $20 million contract for EO. The company was able to field 550 mercenaries altogether, tasked with training and advising government forces, and they also operated a range of major weapons systems—such as armored vehicles, artillery, MI-8 and MI-17 helicopters, and MiG 23 fighter jets—in Angola between 1994 and 1996.[38] The performance of EO in Angola was so impressive that the UN even considered hiring it to intervene in the Rwandan genocide in 1994 but then decided that the world was not yet ready for this step.[39]

The 2003 Iraq War led to another major expansion of the private military industry owing to the decision of the George W. Bush administration to privatize security in Iraq to cut down the number of U.S. troops that needed to be stationed in the country. Companies such as Blackwater, Triple Canopy, and Aegis Defence Services deployed hundreds of heavily armed security contractors. By 2007, the private security contractors in Iraq numbered 48,000, according to a government report.[40] McFate suggested that, in 2010, there were 11,610 private military company (PMC) personnel in Iraq and 14,439 PMC personnel in Afghanistan.[41] The U.S. military and U.S. Department of State continue to use private security contractors in support of U.S. foreign policy.[42]

Even the Ukraine–Russia conflict, which started in February 2022, has involved large paramilitary and mercenary formations such as the Ukrainian Azov Battalion and the Russian Wagner Group. Azov emerged in 2014 and has been described as a right-wing volunteer force; in early 2022, it numbered about 10,000 fighters, of which 95–98 percent were Ukrainian and the rest foreign.[43] In June 2023, there were two Azov brigades that were incorporated into the Ukrainian armed forces but were known for sporting Nazi insignia and enjoyed 'enormous autonomy.'[44] Western and even Japanese mercenaries have been spotted fighting for Ukraine, while Russia has relied extensively on the Wagner Group mercenaries. Wagner is rumored to have been founded by the Russian military intelligence service GRU in 2010, and it functions as Russia's foreign legion.[45] Wagner reportedly fielded up to 50,000 fighters in Ukraine, many of whom were recruited from Russian prisons.[46]

Unconventional Wars

The unconventional nature of wars in developing countries has created increasing challenges for modern armed forces, which are primarily trained and equipped to fight conventional wars against similarly trained and equipped adversaries. Van Creveld claimed that the

> shift from conventional war to low intensity conflict will cause many of today's weapons systems, including specifically those that are most powerful and most advanced, to be assigned to the scrap-heap. Very likely it also will put an end to large-scale military technological research and development as we understand it today.[47]

Although van Creveld was clearly wrong with his assertion that technologically advanced weaponry would become obsolete in the face of the (d)evolution of warfare, he was at least partially right that stealth bombers and main battle tanks are not useful for fighting insurgents and terrorists hiding among populations.

Shadow Wars

The U.S. government has been engaged in a practice of warfare commonly described as 'shadow wars'—quasi-secret or unacknowledged conflicts between the security elements of nation states and a variety of nonstate actors such as terror groups, criminal syndicates, and warlords—conducted under the broad mandates of a 'war on drugs' or a 'war on terror.' Sean McFate suggested that 'shadow wars will dominate' warfare and that '[s]hadow wars are armed conflicts in which plausible deniability, not firepower, forms the center of gravity.'[48]

Shadow wars are not typical cases of covert action, military operations, intelligence operations, or law enforcement, but are often a combination of all of them. In the 1980s and 1990s, the CIA and Special Operations Forces (SOF) conducted a shadow war against the Colombian drug cartels, which included intelligence and law enforcement cooperation with foreign authorities, drug interdiction missions, drug eradication missions, SOF on the ground, sponsoring proxy forces, sabotage, and assassinations.[49] Even before 9/11, the United States had been conducting a shadow war against terror groups, including the PLO, Hezbollah, the Abu Nidal Organization, and Al-Qaeda, which involved intelligence operations, limited military strikes, and a few cases of extraordinary rendition.[50]

After 9/11, counterterrorism became the primary mission of the CIA, which included, in the years 2001–2005, the extraordinary rendition of 119 individuals who were held at black sites, at least 540 drone strikes during the Obama administration, and an unprecedented surge in special operations around the world, with U.S. SOF operating in over 120 countries by 2012.[51] U.S. SOF have participated in a number of armed conflicts in the Middle East and Africa, most notably in Somalia, Syria, and Yemen. These shadow wars have become so routine that they are hardly reported in the Western press anymore, and sometimes even Congress seems to have been left in the dark about some of the highly classified clandestine military operations around the world.[52]

Global Insurgency

Samuel P. Huntington warned in the early 1990s of a coming clash of civilizations.[53] The argument was first outlined in an influential article in *Foreign Affairs* in 1993 and later turned into a book.[54] In his article, Huntington suggested that the Western-dominated international system based on nation states was coming to an end, and that

> the fundamental source of conflict in this new world will not be primarily ideological or primarily economic. The great divisions among humankind and the dominating source of conflict will be cultural. Nation states will remain the most powerful actors in world affairs, but the principal conflicts of global politics will occur between nations and groups of different civilizations.[55]

Shortly after the 9/11 attacks, Huntington declared the 'age of Muslim wars,' noting that two-thirds of all conflicts in 2000 involved Muslims, although Muslims made up only a fifth of the world's population.[56] Culture and religion have again become a source of conflict. As Lind pointed out, 'invasion by immigration can be at least as dangerous as invasion by a state army.'[57]

Al-Qaeda is a terror group that was formed in 1988 by the Palestinian Abdullah Azzam and the Saudi son of a billionaire Osama bin Laden; it had grown out of the Afghan Services Bureau established in 1984.[58] Azzam and bin Laden were organizing and leading a group of foreign fighters motivated by jihadism, who fought against the Soviets and the pro-Soviet regime in Afghanistan.[59] When the conflict was winding down, bin Laden sent his fighters into numerous other conflicts involving Muslims, in Somalia, Bosnia, Chechnya, and Kosovo. In 1996, bin Laden issued a fatwa in which he denounced the occupation of Saudi Arabia by 'crusaders.' In 1998, he made second declaration of war that stated:

> The judgment to kill and fight Americans is an obligation for every Muslim who is able to do so in any country . . . we call upon every Muslim, who believes in Allah to abide by Allah's order by killing Americans and stealing their money anywhere, anytime, and wherever possible.[60]

After 9/11, Al-Qaeda's main base of operations in Afghanistan was destroyed, which resulted in the group being spread out over 40 countries in a loose network of Islamic terror groups. David Kilcullen argued that Al-Qaeda intended to reestablish a global caliphate by spreading jihadist propaganda and unifying Muslim movements around the world.[61] According to Kilcullen,

> the global jihad is clearly an insurgency—a popular movement that seeks to change the status quo through violence and subversion. But whereas traditional insurgencies sought to overthrow established governments or social orders in one state or district, this insurgency seeks to transform the entire Islamic world and remake its relationship with the rest of the globe.[62]

As a result, the West was forced into fighting a global counterinsurgency campaign aimed at defeating the global jihadist insurgency.

Hybrid Warfare

Hybrid warfare (HW) is a term that found its way into the strategic literature and policy documents in the late 2000s and is closely associated with 4GW.[63] There are many different definitions and uses of the term, which has led to some confusion as to what it actually denotes.[64] The term 'hybrid threats' was first introduced by Frank Hoffman in 2007 in a strategy paper for the Potomac Institute, where he suggested that '[h]ybrid wars incorporate a range of different modes of warfare, including conventional capabilities, irregular tactics and formations, terrorist acts including indiscriminate violence and coercion and criminal disorder.'[65] The new and much more complex form of conflict would blur regular and irregular conflict, would aim 'to achieve synergistic effects in the physical and psychological dimensions of conflict,' and would be specifically designed to attack U.S. vulnerabilities.[66]

Hoffman suggested that

> future adversaries (states, state-sponsored groups, or self-funded actors) will exploit access to modern military capabilities including encrypted command systems, man-portable air to surface missiles, and other modern lethal systems, as well as promote protracted insurgencies that employ ambushes, improvised explosive devices (IEDs), and coercive assassinations.[67]

Hoffman's primary example is Hezbollah in Lebanon, which fought a war with Israel in 2006. Hezbollah was able to successfully use irregular tactics in combination with advanced weaponry such as armor-piercing missiles and unmanned aerial vehicles (UAVs) against Israel's conventional forces. Hoffman pointed out that '[t]he battle for perception was just as critical as the strategic strike competition,' as Hezbollah created the (false) perception of having achieved a victory over Israel.[68]

Hoffman distinguished between compound wars and hybrid wars by suggesting that compound wars are major wars that have both regular and irregular operations that synergistically support each other, while hybrid wars would be characterized by an 'operational fusion of conventional and irregular capabilities.'[69] Hoffman connected the concept of hybrid wars with the older notion of 4GW, but pointed at the various analytical shortcomings of the concept, such as an inadequate understanding of military history and the misleading claim that it would amount to an entirely new form of warfare.[70] The way Hoffman treats HW is as a novel type of armed conflict that simply combines advanced technology with irregular tactics and integrates them to great effect.

The Theory of Generational Warfare

The previous sections established that conventional warfare, or Clausewitzian war, is in retreat and is being replaced by forms of armed conflict that are much more complex and messier. This section will explain the basic framework for the idea of generational warfare. Lind et al. postulated that there are successive generations of warfare representing qualitative shifts that make previous forms of military power obsolete. According to Lind, warfare evolved from a focus on massed manpower (1GW—mid-17th century to early 20th century) to a focus on firepower or attrition (2GW—early 20th century to mid-20th century) and, eventually, to a focus on maneuver (3GW – mid to late 20th century).

The first generation of warfare

> reflect[ed] the tactics of the era of the smoothbore musket, the tactics of line and column. These tactics were developed partially in response to technological factors . . . and partially in response to social conditions and ideas . . . [such as] the elan of the revolution and the low training levels of conscripted troops.[71]

The nationalism of the revolution allowed the conscription of large numbers of soldiers who could be effective on the battlefield as new weapons technology made it easier to train soldiers.

The second generation of warfare is seen as 'a response to the rifled musket, breechloaders, barbed wire, the machine gun, and indirect fire. Tactics were based on firepower and movement, and they remained essentially linear . . . Massed firepower replaced massed manpower.'[72] In other words, the second generation was enabled by technological progress that allowed indirect fire by massed artillery to deny enemy forces territory. The major battles of the First World War on the Western Front, which were fought primarily with massed rapid-firing artillery against strong defensive positions, are characteristic of 2GW.

The third generation is seen also as a response to the previous generation's paradigm of massed firepower and it was less of a technological revolution and more of an idea-driven revolution. 'Based on maneuver rather than attrition, third generation tactics were the first truly non-linear tactics. The attack relied on infiltration to bypass and collapse the enemy's combat forces rather than seeking to close with and destroy them.'[73] The experience of the deadlock of the First World War, where large armies fought over small stretches of territory in order to break through defensive lines, resulted in search for a solution, which was maneuver warfare or blitzkrieg. Instead of seeking to destroy the enemy's forces head on in the field, the armed forces utilized speed to bypass them in order to take major objectives that made a continuation of war impossible. The paradigmatic example of 3GW is, hence, the blitzkrieg waged by the Nazis and Japanese in the early years of the Second World War.

This means that 4GW theory sets the late 18th century as the starting point for successive revolutions in war. This Napoleonic era was the high point of the Clausewitzian paradigm of war. The theory claims that the revolutions in war, from the first to the third, were influenced by technological changes, broad societal changes, and new ideas or concepts. The next generation of warfare would always be a response to the previous generation, and it would render the mode of war of the previous generation ineffective and, hence, would make it obsolete. In other words, there is a proclaimed dialectical process where the next generation of warfare is qualitatively different and more advanced than the previous one. Lind claimed that

> [e]ach generational change has been marked by greater dispersion on the battlefield. The fourth generation battlefield is likely to include the whole of the enemy's society. Such dispersion, coupled with what seems likely to be increased importance for actions by very small groups of combatants, will require even the lowest level to operate flexibly on the basis of the commander's intent.[74]

The way Lind et al. initially conceptualized generational warfare was as a sequential process that only goes in one direction, from less complex modes of war to more complex modes of war. This meant that, when 4GW emerged, it was the most advanced or most complex mode of war. Hammes similarly suggested that '[e]ach succeeding generation reached deeper into the enemy's territory in an effort to defeat him.'[75]

New Features of 4GW

When Lind et al. first put forward the concept of 4GW in 1989, it remained largely undefined and speculative. They suggested that there could be a technologically driven fourth generation, which considers emerging technologies as a force multiplier. They stated: 'Technologically it is possible that a very few soldiers could have the same battlefield effect as a current brigade.'[76] These ideas later became known as the Revolution in Military Affairs school of thought, which was first articulated by Andrew Krepinevich and the DoD Office of Net Assessments in 1992 and dominated U.S. military thinking in the 1990s.[77] The DoD's Joint Vision 2010 explored key ideas of the RMA in four sections titled 'Dominant Maneuver,' 'Precision Engagement,' 'Full Dimensional Protection,' and 'Focused Logistics.'[78] The main buzzwords of this era were 'network-centric warfare' and the 'system of systems,' or the idea that advanced ISR capabilities combined with a free flow of information across all military units could enable more complex maneuvers and allow the military to 'adapt to a highly complex situation faster and better than the opponent can.'[79]

The alternative to this vision of future war was an idea-driven fourth generation that considered terrorism to be a potential new mode of war that could blunt or bypass U.S. military superiority.[80] The 4GW school soon questioned the utility of technology in the 'new wars' of the 1990s and, hence, became focused on irregular warfare and conflict short of war. The defining idea of 4GW became that war was reversing to the pre-Westphalian modes of warfare that were characterized by the use of mercenaries, irregular forces, 'bribery, assassination, treachery, betrayal, and even dynastic marriage.'[81] According to Lind and Thiele, '[a]t the heart of this phenomenon, Fourth Generation War, lies not a military evolution, but a political, social, and moral revolution: a crisis of legitimacy of the state.'[82] Fourth generation warfare was supposed to be the latest (and perhaps last) generation of warfare, which was focused on insurgency.

According to Hammes,

[f]ourth generation war actually made its first appearance before World War II . . . Its first practitioner to both write about and successfully execute a concept of 4GW, Mao Tse-Tung, was a product of the intense turmoil that characterized China in the twentieth century.[83]

Mao developed the concept of the people's war, which was about building broad political support for the Chinese Revolution. This meant that particular emphasis was given to the idea that the insurgents treated the Chinese civilian population well, so that they would be more sympathetic towards the goals of the revolutionaries. Hence, Mao relied on both a military and a political strategy for achieving the revolution.

Hammes offered the following definition of 4GW:

Fourth generation war uses all available networks—political, economic, social and military—to convince the enemy's political decision makers that their strategic goals are either unachievable or too costly for the perceived benefit. It is rooted in the fundamental precept that superior political will, when properly employed, can defeat greater economic and military power. 4GW does not attempt to win by defeating the enemy's military forces.

Instead, combining guerrilla tactics or civil disobedience with the soft networks of social, cultural and economic ties, disinformation campaigns and
innovative political activity, it directly attacks the enemy's political will.[84]

Based on the writings of Lind, van Creveld, and Hammes, the following characteristics are associated with 4GW:

- *Asymmetry/insurgency*: Fourth generation is generally a conflict between the
  regular armed forces of a state against irregular nonstate forces. Obviously, insurgency is not a new mode of war as it can be traced back to ancient times. However, 4GW theorists consider 4GW to be an 'evolved insurgency' that has novel
  features which were not present in insurgencies that predate 4GW.[85] According to
  Tim Benbow, the term asymmetry 'means fighting an opponent by using forces,
  tactics, or strategies that are dissimilar.'[86] Asymmetry can either exist in the form
  of a decisive technological advantage over the other, or in the militarily weaker
  side relying on a different mode of conflict that negates the military advantages
  of the other side. It is the claim of 4GW proponents that the new insurgencies are
  particularly difficult to counter with regular armed forces.
- *Attack on the political will*: Unlike earlier insurgencies, 4GW insurgencies are
  more focused on attacking the political will of the adversary and on influencing the adversary's decision-making than on defeating the adversary's military
  forces.[87] The insurgents seek to gain the moral high ground and to undermine
  the legitimacy of their state adversaries, which is more decisive than the military
  means employed by 4GW adversaries. As Lind and Thiele pointed out: '*what
  works for you on the physical (and sometimes mental) level often works against
  you at the moral level*. It is therefore very easy to win all the tactical engage-
  ments in a Fourth Generation Conflict yet still lose the war.'[88]
- *Blurring of war and peace*: 4GW blurs the distinction between war and peace,
  as hostilities are sporadic and there is no definable battlefield. Conflicts may ex-
  tend over years and decades with no clear outcome. When confronted with 4GW
  adversaries, state militaries may be reluctant to call these conflicts war, which
  will make it difficult to draw a line between peacetime and wartime military
  activities. Within the larger context of 4GW, there are also so-called 'shadow
  wars' or quasi-secret or unacknowledged and open-ended conflicts between the
  security elements of nation states and a variety of nonstate actors such as terror
  groups, criminal syndicates, and warlords, which are conducted under the broad
  mandate of a 'war on drugs' or a 'war on terror.'
- *New belligerents*: During the 1990s, there was a reemergence of mercenaries
  and private security companies as significant actors in international security.[89]
  The South African mercenary outfit Executive Outcomes was hired by several
  governments to intervene in internal conflicts, most notably in Angola, Sierra
  Leone, and Papua New Guinea. The 'new condottieri' could increasingly privatize armed conflict and could augment, if not replace altogether, regular armed
  forces fighting in the 'new wars.'[90] Sean McFate suggested that '[s]tate erosion
  encourages new kinds of global powers. The vacuum left by retreating states

will be filled by insurgents, caliphates, corporatocracies, narco-states, warlord kingdoms, mercenary overlords, and wastelands.'[91] States become actors among many others in a much more complex security environment.

- *Blurring of the distinction between combatant and non-combatant*: A major feature of 4GW is the expanded role of civilians in insurgencies, which has made it increasingly difficult to clearly distinguish between combatants and non-combatants or military and civilian targets. As Lind et al. observed, '[t]he distinction between "civilian" and "military" may disappear.'[92] Fourth generation warfare adversaries can exploit this by attacking the legitimacy of regular forces when they target civilians who assist the insurgents in more indirect ways or when civilians are accidentally targeted owing to the difficulty of clearly distinguishing who is participating in hostilities and who is not. More ruthless belligerents have used civilians as human shields in order to create moral and political dilemmas for modern armed forces. Ralph Peters stated that

  [w]e will face opponents for whom treachery is routine, and they will not be impressed by tepid shows of force with restrictive rules of engagement. Are we able to engage in and sustain the level of sheer violence it can take to eradicate this kind of threat? To date, the Somalia experience says 'No.'[93]

- *Networks*: Fourth generation warfare opponents do not have a traditional hierarchical organization and instead rely on a network organization that is optimized for attacking the minds of military decision-makers.[94] These networks reach deep within society and enable them to use different types of power and influence, including political, economic, and the media. Network organizations are also more flexible, adaptive, and resilient. Networks are important for the political mobilization that is necessary for sustaining and winning fourth generation conflicts. They may also include NGOs, protest movements, and other political activist groups and they may function as fronts for militant groups.

- *Lack of infrastructure*: Fourth generation warfare opponents generally lack a traditional infrastructure for fighting wars that can be attacked, such as factories, military bases, or large weapons systems.[95] Their primary infrastructure is their networks consisting of key personnel. Fourth generation warfare opponents also lack other hard assets that can be targeted, which makes it difficult for state militaries to defeat them. The only way of militarily attacking 4GW opponents is by way of precision attacks aimed at disrupting human networks, which is referred to in a counterterrorism framework as extraordinary rendition and targeted killing.

- *The erosion of the state monopoly on violence*: The new 4GW adversaries challenge the monopoly of violence of the state that was established after 1648. Hence, '4GW is a throwback to warfare before the Peace of Westphalia.'[96] The inability of states to defeat 4GW nonstate adversaries quickly means that the state monopoly of violence gets eroded over time. Nonstate actors may be able to control a portion of a state's territory, or they may be able to establish competing structures for administrative control of territories. Fourth generation wars tend to be long-lasting conflicts and, the longer they continue, the more they destroy state institutions and the ability of governments to govern.

Fighting 4GWs

Hammes developed a strategy for defeating 4GW opponents. First of all, Hammes cautioned that technology may not be decisive in 4GW, although it may provide various tactical advantages. More important would be organizational and doctrinal changes in taking on 4GW opponents.

- *Leveraging all elements of national power*: Fourth generation warfare seeks to leverage many different sources of power and influence in addition to military power. In order to defeat 4GW adversaries, it is therefore necessary to bring to bear all available elements of national power. In practice, this means better interagency cooperation across governments and a reduced role for the military and military force.
- *Organizational and cultural changes*: As 4GW adversaries use a network form of organization, they can be more flexible, more adaptive, and resilient, which makes it much harder for traditional hierarchical organizations to fight them. 'It takes a network to fight a network.'[97] As a result, states and state militaries need to function more like networks and they need to become flatter.[98] This requires a major cultural change within governments and major government bureaucracies such as the armed forces.
- *Global or transnational perspective*: Fourth generation warfare adversaries understand that they operate within an international or global environment, and that their victory depends not only on what happens within the country in which they fight, but also on outside support and activities outside their country. Winning the support of more powerful outside actors, such as rival states or groups, could be key to their success. The homeland is no longer a sanctuary, and the 4GW opponent may have to be fought at home and not just abroad.[99]
- *Focus on people rather than technology*: Owing to substantial technological advantages that the United States has enjoyed over its adversaries for many decades, there is a tendency to focus on technological solutions rather than on other important factors. Hammes argued: 'What really matters [in 4GW] are well-trained, intelligent, creative people guided by a coherent long-term strategy.'[100] This means that success in fighting 4GW opponents requires a greater emphasis on training and retaining highly skilled personnel who have sufficient understanding of the cultures and societies in which they have to fight.
- *Focus on the moral dimension rather than on the physical or mental dimension*: Since the objective of 4GW opponents is to undermine the legitimacy of their state adversaries, it becomes more important to occupy the moral high ground than to win battles. This means that it is better to capture adversaries than to kill them with stand-off weaponry, even if this is physically much harder to accomplish.[101] It becomes key to victory to deny adversaries a compelling narrative that legitimizes their fight.

- *Silent warfare*: Silent war can be defined as 'waging war where the war, political desires, combatants, and the strategic forms of power used in the war are invisible, not very energetic, and lean towards influence.'[102] Lind suggested that modern militaries should adopt the Mafia model when dealing with 4GW opponents.

> One key to a mafia's success is the concealed use of force. When an individual needs to be 'whacked,' then it is usually done with little fanfare and in the shadows. The rule is 'no fingerprints' Unless there is a specific message intended for a larger audience, people killed by the Mafia are seldom found.[103]

   This means that, in 4GW combat, actions may occur covertly and with no attribution or declaration of responsibility.

The 4GW school was quite influential in the 1990s and early 2000s owing to the increase in civil conflicts and also in Western military interventions in these conflicts. The decade from 2001 to 2011 was shaped by the War on Terror, which meant that the U.S. government conducted global counterterrorism operations and fought two major counterinsurgency campaigns, in Afghanistan and Iraq. Many ideas of the 4GW eventually found their way into U.S. counterinsurgency doctrine, which can be considered its lasting impact or legacy.[104]

Criticism of the 4GW School

The reception of the 4GW school has been mixed and has been dominated by more critical voices, which may explain why there has been far less professional interest, academic or military, in this school of thought since the 1990s. Before continuing with 5GW, it is therefore important to consider some of the criticism of 4GW and also whether that criticism also tarnishes the 5GW school. The main criticism that is often made in relation to novel theories is the claim that the ideas lack novelty or originality, and that they merely amount to a repackaging of fairly established ideas.
   Military historian Antulio Echevarria heavily criticized the 4GW school in his study *Fourth-Generation War and Other Myths*.[105] One of his criticisms was that 4GW amounted to an 'empty theory' that was 'founded on myths about the so-called Westphalian system and the theory of *blitzkrieg*.'[106] Various critics have pointed out the following flaws of 4GW:

- *Poor understanding of history*: Echevarria suggested that 4GW theorists' understanding of the Westphalian Treaty or of warfare in the 20th century was overly simplistic, if not severely flawed. Trinitarian war never existed in the way envisioned by van Creveld, making the idea of its replacement with nontrinitarian war meaningless. According to Echevarria,

> those who, like van Creveld, refer to Clausewitz' trinity as people, military, and government, do not understand it. Strictly speaking, there is no such

thing as trinitarian war, because the three forces are present in every war, not just wars of nation states.[107]

Furthermore, none of the generations of war proclaimed by the 4GW school existed in a pure form, as there were always other modes of war that existed alongside each generation. The primary example of 2GW is the First World War, which was supposed to be all about firepower and attrition. This is easily contradicted by other dominant aspects of this war involving attacks on the political will of the people, such as bombings of cities from the air or the naval blockade that starved the German people. Insurgency and HW existed as modes of conflict long before the supposed arrival of 4GW in the 1930s.[108]

- *Weaknesses of the theory*: Echevarria has heavily criticized the theoretical framework of 4GW, calling it 'unnecessary' and saying that 'it only serves to undermine the credibility of those who employ it in the hope of inspiring the right kinds of change.'[109] The theory suggests that war evolves in the way of a linear succession of generations, with each generation of war being more evolved than the previous one and each new generation displacing the previous one.[110] The theory is, hence, unable to explain the coexistence of different generations of war.[111] More importantly, the theory fails to account for the role of other instruments of power than military force in the conduct of modern war, such as the 'integration of political, economic, and social power.'[112] By proclaiming a logical succession of manpower, firepower, maneuver, and insurgency, the 4GW proponents failed to understand the difference between 'military means or techniques' (manpower, firepower, maneuver) and a different 'form of war' (insurgency).[113] 4GW was also a patchwork of different ideas, and there is a clear tension between the generational framework of warfare and van Creveld's claim that future war would be largely of the 'nontrinitarian' variety. In short, within van Creveld's own theory, there is little to suggest that nontrinitarian war would itself be merely a temporary generation rather than a final trajectory for modern warfare. Traditionalists have also argued that warfare has not evolved beyond some tactical or operational innovations, and that the principles of strategy are basically eternal and, hence, universally applicable to all wars throughout history. This would make it unnecessary and misleading to conjure up new generations of warfare since war can change some of its characteristics but cannot change its nature.[114] Obviously, this traditionalist criticism applies not only to the 4GW and 5GW theories, but to *all* theories that argue that there are evolutionary or revolutionary changes in warfare.
- *Inadequate guidance for the future*: The final concern questions the relevance of 4GW for even understanding the current reality of war and conflict, which does not correspond to the earlier predictions of 4GW, or doing justice to the ability of regular armed forces to deal with asymmetric threats. Since insurgency is not a new form of war, the challenge of regular armed forces fighting insurgencies is equally not a new problem.[115] The 4GW school was too quick to declare the irrelevance of state militaries and of advanced technology. The new wars would be both high-tech and low-tech. It would be not accurate to

suggest that advanced military technology would not make much of a difference or that 4GW opponents would always have an advantage over better-trained and better-equipped regular armed forces just by prioritizing the political dimension of a conflict. In fact, 4GW opponents in Vietnam, Iraq, and Afghanistan almost always lost their battles with the U.S. military. Hence, it would appear that 4GW proponents have exaggerated the threat this mode of war actually poses to regular armed forces and their ability to win wars.[116] Wirtz argued that 4GW 'is not combat effective against a modern combined-arms military. But as Hammes correctly notes, it does pose a political challenge.'[117] The idea that future warfare could be narrowly defined as 'evolved insurgency' and made to fit into this framework is also questionable.[118] Edward Luttwak correctly suggested that '[t]o refocus military resources on "Fourth Generation Warfare" would be especially unfortunate for the United States' in view of the rise of China and the possible return of great power conflict.[119]

The Reemergence of Conventional War

Something few analysts predicted in the 1990s and early 2000s is the reemergence of great power conflict since 2014 and of major conventional war since 2022. In 2014, the pro-Russian government of Viktor Yanukovych was overthrown in Ukraine, following severe rioting and police violence in Kyiv, which triggered a covert Russian intervention in Eastern Ukraine and the Russian annexation of Crimea.[120] The tensions between Russia and the West worsened after the shooting down of the Malaysian airliner MH17 over Eastern Ukraine, supposedly by Ukrainian separatists using a Russian-supplied air defense system, in July 2014.[121] During the September 2014 NATO summit in Wales, NATO members pledged to spend 2 percent of their GDP on defense so that the alliance would be better equipped to counter a revanchist Russia. NATO developed its Readiness Action Plan, which included NATO maneuvers and deployments on Russia's periphery, as well as the establishment of a Very High Readiness Joint Task Force, composed of several thousand alliance troops, that is deployable within days.[122] The Minsk Protocol of September 2014 established a ceasefire and required Russia to withdraw its mercenaries and equipment from Eastern Ukraine. Unfortunately, the Minsk agreement did not end the conflict in Ukraine.

In February 2022, Russia carried out a 'Special Military Operation' in Ukraine, which was an incursion of Russian forces into Ukrainian territory with the stated objective of 'de-Nazifying' and 'de-militarizing' Ukraine and to respond to the '"threatening expansion" of NATO towards Russia's borders.'[123] Russia sent over 150,000 soldiers into Ukraine, where they faced a more formidable Ukrainian defense than expected.[124] The conflict turned into a major conventional war, which, at time of writing, had cost the lives of tens of thousands of Ukrainians and Russians.[125] NATO countries promised Ukraine support in the form of financial and military aid to repel the Russian forces.

As of August 2022, the U.S. government had delivered 16 HIMARS artillery systems, 1,500 TOW missiles, 126 155-mm howitzers with over 800,000 rounds,

20 120-mm mortars with 85,000 rounds, eight NASAMS air defense systems, 1,400 micro drones, 20 Mi-17 helicopters, 1,400 Stinger missiles, 8,500 Javelin anti-tank missiles, and 27,000 other anti-tank systems.[126] The Royal United Services Institute estimated that Russia was firing 7,176 artillery rounds per day in May 2022, which would mean that the Ukrainians needed to have a similar capacity to keep up. The Western industrial defense base is not currently ready for war on an industrial scale, as it cannot produce the required amounts of munitions.[127] The global market for major weapons systems such as tanks, fighter jets, and warships has been rapidly growing since 2022, with a potential war between NATO and Russia, as well as another potential war between the United States and China over Taiwan, on the horizon. This raises the question of whether the world will return to the era of great power wars rather than moving in the direction of a post-Westphalian world described by 4GW theorists.

Although conventional warfare again preoccupies the military establishments of the great powers, there are many reasons as to why 4GW remains relevant. First of all, the fundamental strategic situation that characterized the original Cold War has not substantially changed: The great powers are still constrained by the risk of escalation to all-out nuclear war. It makes no sense to assume that a major conventional war between great powers would not eventually escalate to a nuclear war, which all sides would very likely prefer to avoid. President Biden made it clear in a March 2022 speech: 'We will not fight a war against Russia in Ukraine. Direct conflict between NATO and Russia is World War III, something we must strive to prevent.'[128]

Second, the Ukraine War will end eventually and will likely leave Eastern Europe substantially less politically stable and less prosperous. The main security threats for Western countries will be internal and transnational rather than conventional military invasion. Major Western countries such as the United States may already be in a pre-civil war state, as argued by Lind:

> Mobs loot, burn, and vandalize while politicians advocate defunding the police. A commune was established in Seattle and turned into *Lord of the Flies* while government did nothing. Blacks demand equal treatment from police despite a violent crime rate many times greater than that of whites, and mainstream media will not report honestly the differences in crime rates. 'Wokeness' spreads among idle youth who flunked English 101. What is going on? What is going on, right here on American soil, is war; a new kind of war that is also very old, waged by entities other than states. I call it Fourth Generation War.[129]

Conclusion

Fourth generation warfare proponents believe that both the nature and the character of war are undergoing a fundamental change because of the decline of the nation state and the end of the Westphalian international system. The reasons for the decline of the nation state are complex, but relate mostly to globalization and technology, which have made borders more porous and have empowered nonstate actors. National sovereignty has been eroded by the creation of international organizations with supranational features that result in overlapping authority and much greater political complexity. Terror groups, insurgents, warlords, militias, and criminal organizations now control territory in some parts of the world and, hence, are able to again compete with governments in the domain of warfare. Most conflicts are intrastate conflicts characterized by insurgencies, terrorism, and political instability. Conventional militaries have not been very successful in intervening in such wars. Fourth generation warfare proponents are generally pessimistic about the ability of modern militaries to leverage their strengths in terms of superior training, organization, and technology, as 4GW adversaries mostly fight in the moral and political arena and lack any tangible center of gravity that can be attacked. Therefore, 4GW proponents have suggested focusing on the moral and political dimensions of a conflict since a moral defeat would be more decisive than a physical defeat. It is hard to dismiss the continuing relevance of many ideas of the 4GW school in today's world, despite the recent resurgence of conventional war in Ukraine and the looming Chinese invasion of Taiwan.

Notes

1 William S. Lind and Gregory A. Thiele, *4th Generation Warfare Handbook* (Kouvola, Finland: Castalia House, 2015); Martin van Creveld, *The Transformation of War* (New York: Free Press, 1991); and Thomas X. Hammes, *The Sling and the Stone: On War in the 21st Century* (St. Paul, MN: Zenith Press, 2004).
2 Antulio Echevarria, *Fourth-Generation War and Other Myths* (Carlisle Barracks, PA: Strategic Studies Institute, November 2005), 2–9.
3 Antulio Echevarria, 'Deconstructing the Theory of Fourth-Generation Warfare,' in Terry Terriff, Aaron Karp, and Regina Karp, *Global Insurgency and the Future of Armed Conflict: Debating Fourth-Generation Warfare* (London: Routledge, 2007), 59.
4 Tim Benbow, *The Magic Bullet: Understanding the Revolution in Military Affairs* (London: Brassey's, 2004), 154.
5 Martin van Creveld, *The Rise and Decline of the Sate* (Cambridge University Press, 1999), 1.
6 Van Creveld, *The Rise and Decline of the Sate*, 147–149.
7 Herfried Münkler, *The New Wars* (Cambridge: Polity, 2004), 58–59.
8 Janice E. Thomson, *Mercenaries, Pirates, and Sovereigns* (Princeton, NJ: Princeton University Press, 1994).
9 Van Creveld, *The Transformation of War*, 40.
10 Van Creveld, *The Transformation of War*, 40.
11 Van Creveld, *The Transformation of War*, 5–7.
12 Francis Fukuyama, *The End of History and the Last Man* (New York: Free Press, 1992), 262.
13 Steven Metz, *Rethinking Insurgency* (Carlisle, PA: U.S. Army War College Press, 2007), 10.
14 Van Creveld, *The Transformation of War*, 20.
15 Carl Schmitt, *Theory of the Partisan: Intermediate Commentary on the Concept of the Political* (New York: Telos Press, 2007), 4.
16 John Arquilla, *Insurgents, Raiders, and Bandits: How Masters of Irregular Warfare Shaped Our World* (Chicago, IL: Ivan R. Dee, 2011), 42.

17 Robert Taber, *War of the Flea: The Classic Study of Guerrilla Warfare* (Lincoln, NE: Potomac Books, 2002), 2–3.
18 Quoted in Gerard J. DeGroot, *A Noble Cause? America and the Vietnam War* (New York: Pearson, 2000), 104.
19 Taber, *War of the Flea*, 40.
20 Schmitt, *Theory of the Partisan*, 30.
21 Schmitt, *Theory of the Partisan*, 37.
22 Sean McFate, *The Modern Mercenary: Private Armies and What They Mean for World Order* (Oxford: Oxford University Press, 2014), 73.
23 McFate, *The Modern Mercenary*, 73.
24 Philip Cerny, 'Neomedievalism, Civil War and the New Security Dilemma: Globalisation as Durable Disorder,' *Civil Wars* 1, no. 1 (1998): 39.
25 McFate, *The Modern Mercenary*, 75.
26 McFate, *The Modern Mercenary*, 77.
27 McFate, *The Modern Mercenary*, 80.
28 McFate, *The Modern Mercenary*, 82.
29 Mary Kaldor, *New and Old Wars: Organized Violence in a Global Era* (Cambridge: Polity, 2001), 102–103.
30 Robert Kaplan, 'The Coming Anarchy,' *The Atlantic*, February 1994.
31 Kaplan, 'The Coming Anarchy.'
32 Ralph Peters, 'The New Warrior Class,' *Parameters* 24, no. 1 (1994): 16–26.
33 Steven Metz and James Kievit, *The Revolution in Military Affairs and Conflicts Short of War* (U.S. Army War College, Strategic Studies Institute, 1994), 2.
34 Organization of the African Union, 'OAU Convention for the Elimination of Mercenarism in Africa,' Libreville, Gabon, July 3, 1977.
35 Peter Singer, *Corporate Warriors: The Rise of the Private Military Industry* (Ithaca, NY: Cornell University Press, 2003), 45.
36 Singer, *Corporate Warriors*, 49.
37 Singer, *Corporate Warriors*, 49–60.
38 Herbert M. Howe, 'Private Security Forces and African Stability: The Case of Executive Outcomes,' *The Journal of Modern African Studies* 36, no. 2 (1998): 312.
39 William Shawcross, *Deliver Us from Evil: Warlords and Peacekeepers in a World of Endless Conflict* (London: Bloomsbury, 2001), 122–123.
40 Jeremy Scahill, 'Bush's Rent-an-Army,' *Los Angeles Times*, January 25, 2007.
41 McFate, *The Modern Mercenary*, 23.
42 Congressional Research Service, 'Defense Primer: Department of Defense Contractors,' *Congressional Research Service Report*, January 17, 2023, available at: https://crsreports.congress.gov/product/pdf/IF/IF10600
43 Sudarsan Rhagavan, Loveday Morris, Claire Parker, and David L. Stern, 'Right-Wing Azov Battalion Emerges as a Controversial Defender of Ukraine,' *Washington Post*, April 6, 2022.
44 Lev Golinkin, 'The Western Media Is Whitewashing the Azov Battalion,' *The Nation*, June 13, 2023, available at: www.thenation.com/article/world/azov-battalion-neo-nazi/
45 Kimberly Marten, 'Russia's Use of Semi-State Security Forces: The Case of the Wagner Group,' *Post-Soviet Affairs* 35, no. 3 (2019): 187.
46 Nathan Luna, Leah Vredenbregt, and Ivan Pereira, 'What Is the Wagner Group? The "Brutal" Russian Military Unit in Ukraine,' *ABC News*, June 23, 2023, available at:https://abcnews.go.com/International/International/wagner-group-brutal-russian-military-group-fighting-ukraine/story?id=96665326
47 Van Creveld, *The Transformation of War*, 205.
48 Sean McFate, *The New Rules of War: Victory in the Age of Durable Disorder* (New York: William Morrow, 2019), 199.

49 The story of the U.S. shadow war on the Medellin Cartel in the early 1990s is told in Mark Bowden's *Killing Pablo: The Hunt for the Richest, Most Powerful Criminal in History* (London: Atlantic Books, 2001).

50 Timothy Naftali, *Blind Spot: The Secret History of American Counterterrorism* (New York: Basic Books, 2005).

51 U.S. Senate, 'Report of the Senate Select Committee on Intelligence Committee Study of the Central Intelligence Agency's Detention and Interrogation Program,' December 9, 2014, 13; Micah Zenko, 'Obama's Final Drone Strike Data,' *Council on Foreign Relations* (blog), January 20, 2017, available at: www.cfr.org/blog/obamas-final-drone-strike-data; Nick Turse, *The Changing Face of Empire: Special Ops, Drones, Spies, Proxy Fighters, Secret Bases, and Cyberwarfare* (Chicago, IL: Haymarket Books, 2012), 18.

52 Jennifer Kibbe, 'Conducting Shadow Wars,' *Journal of National Security Law & Policy* 5 (2012): 385.

53 Tim Benbow, 'Talking 'Bout Our Generation? Assessing the Concept of "Fourth Generation Warfare,"' *Comparative Strategy* 27, no. 2 (2008): 149.

54 Samuel P. Huntington, *The Clash of Civilizations and the Remaking of World Order* (New York: Simon & Schuster, 1998).

55 Samuel P. Huntington, 'The Clash of Civilizations?' *Foreign Affairs* 72, no. 3 (1993): 22.

56 Samuel P. Huntington, 'The Age of Muslim Wars,' *Newsweek*, 138, no. 25 (2001).

57 William S. Lind, 'Understanding Fourth Generation War,' *Military Review* (September/October 2004): 13.

58 Rohan Gunaratna, *Inside Al Qaeda: Global Network of Terror* (London: Hurst, 2003), 3.

59 John Cooley, *Unholy Wars: Afghanistan, America and International Terrorism* (London: Pluto Press, 2002), 97–98.

60 Peter Bergen, *Holy War, Inc., Inside the Secret World of Osama bin Laden* (London: Phoenix, 2002), 99.

61 David Kilcullen, 'Countering Global Insurgency,' *Journal of Strategic Studies* 28, no. 4 (2005): 599–600.

62 Kilcullen, 'Countering Global Insurgency,' 604.

63 Frank Hoffman, 'Combating Fourth Generation Warfare,' in: Terriff et al., *Global Insurgency and the Future of Armed Conflict*, 178.

64 Chiara Libiseller, '"Hybrid Warfare" as an Academic Fashion,' *Journal of Strategic Studies* 46, no. 4 (2023): 858.

65 Frank Hoffman, *Conflict in the 21st Century: The Rise of Hybrid Wars* (Arlington, VA: Potomac Institute, 2007), 14.

66 Hoffman, *Conflict in the 21st Century*, 7.

67 Hoffman, *Conflict in the 21st Century*, 28.

68 Hoffman, *Conflict in the 21st Century*, 38.

69 Frank Hoffman, 'Hybrid Warfare and Challenges,' *Joint Forces Quarterly* 52, no. 1 (2009): 36.

70 Hoffman, *Conflict in the 21st Century*, 19.

71 Lind et al., 'The Changing Face of War,' 86.

72 Lind et al., 'The Changing Face of War,' 86–87.

73 Lind et al., 'The Changing Face of War,' 86–87.

74 Lind et al., 'The Changing Face of War,' 87.

75 Hammes, *The Sling and the Stone*, 31.

76 Lind et al., 'The Changing Face of War,' 88.

77 Andrew Krepinevich, *The Military Technical Revolution: A Preliminary Assessment* (Washington, DC: Center for Strategic and Budgetary Assessments, 2002).

78 U.S. Department of Defense, 'Joint Vision 2010,' Washington, DC: U.S. Joint Chiefs of Staff (1996).

79 Bill Owens, *Lifting the Fog of War* (New York: Farrar Strauss Giroux, 2000), 118.

80 Lind et al., 'The Changing Face of War,' 88–90.

81 Lind and Thiele, *4th Generation Warfare Handbook*, 6.

82 Lind and Thiele, *4th Generation Warfare Handbook*, 7.
83 Hammes, *The Sling and the Stone*, 44.
84 Thomas Hammes, 'War Evolves into the Fourth Generation,' in Terriff et al., *Global Insurgency and the Future of Armed Conflict*, 42.
85 Hammes, *The Sling and the Stone*, 208.
86 Benbow, *The Magic Bullet*, 154–155.
87 Hammes, 'War Evolves into the Fourth Generation,' 36.
88 Lind and Thiele, *4th Generation Warfare Handbook*, 10; original italics.
89 Thomas Hammes, 'Fourth Generation Warfare Evolves, Fifth Emerges,' *Military Review* 87, no. 3 (2007): 17–18.
90 Eugene Smith, 'The New Condottieri and US Policy: The Privatization of Conflict and Its Implications,' *Parameters* 32, no. 4 (2002): 104–119.
91 McFate, *The New Rules of War*, 149.
92 Lind et al., 'The Changing Face of War,' 88.
93 Peters, 'The New Warrior Class,' 6.
94 Lind and Thiele, *4th Generation Warfare Handbook*, 32.
95 Hammes, *The Sling and the Stone*, 209.
96 Gary Anderson, 'The End of the Peace of Westphalia: Fourth Generation Warfare,' *Small Wars Journal*, October 23, 2013, available at: https://smallwarsjournal.com/jrnl/art/the-end-of-the-peace-of-westphalia-fourth-generation-warfare
97 John Arquilla, 'The End of War as We Knew It? Insurgency, Counterinsurgency and Lessons from the Forgotten History of Early Terror Networks,' *Third World Quarterly* 28, no. 2 (2007): 383.
98 Lind and Thiele, *4th Generation Warfare Handbook*, 32.
99 The 2018 U.S. National Defense Strategy remarkably stated: 'It is now undeniable that the *homeland is no longer a sanctuary*. America is a target, whether from terrorists seeking to attack our citizens; malicious cyber activity against personal, commercial, or government infrastructure; or political and information subversion.' U.S. Department of Defense, 'Summary of the National Defense Strategy of the United States of America,' January 19, 2018, available at: https://dod.defense.gov/Portals/1/Documents/pubs/2018-National-Defense-Strategy-Summary.pdf, 3.
100 Hammes, *The Sling and the Stone*, 232–233.
101 Lind and Thiele, *4th Generation Warfare Handbook*, 12.
102 L.C. Rees, 'The End of the Rainbow: Implications of 5GW for a General Theory of War,' in: Daniel H. Abbott, *The Handbook of 5GW* (Ann Arbor, MI: Nimble Books, 2010), 27.
103 Lind and Thiele, *4th Generation Warfare Handbook*, 49.
104 Michael J. Artelli and Richard F. Deckro, 'Fourth Generation Operations: Principles for the "Long War,"' *Small Wars & Insurgencies* 19, no. 2 (2008): 221–237.
105 Echevarria, *Fourth-Generation War and Other Myths*.
106 Echevarria, *Fourth-Generation War and Other Myths*, v.
107 Echevarria, 'Deconstructing the Theory of Fourth-Generation Warfare,' 60.
108 Peter Mansoor pointed at the Peloponnesian War of the 5th century BCE where the Athenians sponsored incursions and an insurgency in Sparta. Peter Mansoor, 'Introduction: Hybrid Warfare in History,' in: Williamson Murray and Peter R. Mansoor (eds.), *Hybrid Warfare: Fighting Complex Opponents from the Ancient World to the Present* (Cambridge: Cambridge University Press, 2012), 4.
109 Echevarria, 'Deconstructing the Theory of Fourth-Generation Warfare,' 58.
110 Echevarria, 'Deconstructing the Theory of Fourth-Generation Warfare,' 62.
111 Michael Evans, 'Elegant Irrelevance Revisited: A Critique of Fourth Generation Warfare,' in Terriff et al., *Global Insurgency and the Future of Armed Conflict*, 69.
112 Echevarria, *Fourth-Generation War and Other Myths*, 15.
113 Echevarria, *Fourth-Generation War and Other Myths*, 15.
114 Colin S. Gray, *Another Bloody Century: Future Warfare* (London: Weidenfeld & Nicolson, 2005), 371.

115 Rod Thornton, 'Fourth Generation: A "New" Form of Warfare,' in Terriff et al., *Global Insurgency and the Future of Armed Conflict*, 91.
116 James J. Wirtz, 'Politics with Guns: A Response to T.X. Hammes' "War Evolves into the Fourth Generation,"' in Terriff et al., *Global Insurgency and the Future of Armed Conflict*, 47, 49.
117 Ibid., 51.
118 Evans, 'Elegant Irrelevance Revisited,' 72.
119 Edward Luttwak, 'A Brief Note on "Fourth Generation Warfare,"' in Terriff et al., *Global Insurgency and the Future of Armed Conflict*, 52.
120 Mitchell Orenstein, *The Lands Between: Russia vs. the West and the New Politics of Hybrid War* (Oxford: Oxford University Press, 2019), 42.
121 Joshua P. Mulford, 'Non-State Actors in the Russo-Ukrainian War,' *Connections* 15, no. 2 (2016): 90, 102.
122 Orenstein, *The Lands Between*, 73.
123 T.D. Gill, 'The Jus ad Bellum and Russia's "Special Military Operation" in Ukraine,' *Journal of International Peacekeeping* 25 (2022): 122.
124 Bonnie Berkowitz and Artur Galocha, 'Why the Russian Military Is Bogged Down by Logistics in Ukraine,' *Washington Post*, March 30, 2022.
125 Casualty estimates in the war are highly contentious since official casualty numbers tend to be politicized by the belligerent parties. U.S. intelligence documents leaked in April 2023 indicated 354,000 casualties for both sides. Guy Falconbridge, 'Ukraine War, Already with up to 354,000 Casualties, Likely to Last Past 2023—US Documents,' *Reuters*, April 12, 2023, available at: www.reuters.com/world/europe/ukraine-war-already-with-up-354000-casualties-likely-drag-us-documents-2023–04–12/
126 Jordan Williams, 'Here's Every Weapon US Has Supplied to Ukraine with $13 Billion,' *The Hill*, August 26, 2022, available at: https://thehill.com/policy/defense/3597492-heres-every-weapon-us-has-supplied-to-ukraine-with-13-billion/
127 Alex Vershenin, 'The Return of Industrial Warfare,' *RUSI*, June 17, 2022, available at: www.rusi.org/explore-our-research/publications/commentary/return-industrial-warfare
128 Brett Samuels, 'Biden: Direct Conflict between NATO and Russia Would Be "World War III,"' *The Hill*, March 11, 2022, available at: https://thehill.com/policy/international/597842-biden-direct-conflict-between-nato-and-russia-would-be-world-war-iii/
129 William Lind, 'Fourth Generation War Comes to a Theater near You,' *Chronicles Magazine* (October 2020), www.chroniclesmagazine.org/fourth-generation-war-comes-to-a-theater-near-you-1/

References

Anderson, Gary, 'The End of the Peace of Westphalia: Fourth Generation Warfare,' *Small Wars Journal*, October 23, 2013, available at: https://smallwarsjournal.com/jrnl/art/the-end-of-the-peace-of-westphalia-fourth-generation-warfare
Arquilla, John, 'The End of War as We Knew It? Insurgency, Counterinsurgency and Lessons from the Forgotten History of Early Terror Networks,' *Third World Quarterly* 28, no. 2 (2007): 369–386.
———, *Insurgents, Raiders, and Bandits: How Masters of Irregular Warfare Shaped Our World* (Chicago, IL: Ivan R. Dee, 2011).
Artelli, Michael J. and Deckro, Richard F., 'Fourth Generation Operations: Principles for the "Long War,"' *Small Wars & Insurgencies* 19, no. 2 (2008): 221–237.
Benbow, Tim, *The Magic Bullet: Understanding the Revolution in Military Affairs* (London: Brassey's, 2004).

————, 'Talking 'Bout Our Generation? Assessing the Concept of "Fourth Generation Warfare,"' *Comparative Strategy* 27, no. 2 (2008): 148–163.

Bergen, Peter, *Holy War, Inc., Inside the Secret World of Osama bin Laden* (London: Phoenix, 2002).

Berkowitz, Bonnie and Galocha, Artur, 'Why the Russian Military Is Bogged Down by Logistics in Ukraine,' *Washington Post*, March 30, 2022.

Bowden, Mark, *Killing Pablo: The Hunt for the Richest, Most Powerful Criminal in History* (London: Atlantic Books, 2001).

Cerny, Philip, 'Neomedievalism, Civil War and the New Security Dilemma: Globalisation as Durable Disorder,' *Civil Wars* 1, no. 1 (1998): 36–64.

Congressional Research Service, 'Defense Primer: Department of Defense Contractors,' Congressional Research Service Report, January 17, 2023, available at: https://crsreports. congress.gov/product/pdf/IF/IF10600

Cooley, John, *Unholy Wars: Afghanistan, America and International Terrorism* (London: Pluto Press, 2002).

DeGroot, Gerard J., *A Noble Cause? America and the Vietnam War* (New York: Pearson, 2000).

Echevarria, Antulio J., *Fourth-Generation War and Other Myths* (Carlisle Barracks, PA: Strategic Studies Institute, 2005).

————, 'Deconstructing the Theory of Fourth-Generation Warfare,' in Terry Terriff, Aaron Karp, and Regina Karp, *Global Insurgency and the Future of Armed Conflict: Debating Fourth-Generation Warfare* (London: Routledge, 2007), 58–66.

Evans, Michael, 'Elegant Irrelevance Revisited: A Critique of Fourth Generation Warfare,' in Terriff et al., *Global Insurgency and the Future of Armed Conflict*, 67–74.

Falconbridge, Guy, 'Ukraine War, Already with up to 354,000 Casualties, Likely to Last Past 2023—US Documents,' *Reuters*, April 12, 2023, available at: www.reuters. com/world/europe/ukraine-war-already-with-up-354000-casualties-likely-drag-us-documents-2023-04-12/

Fukuyama, Francis, *The End of History and the Last Man* (New York: Free Press, 1992).

Gill, T.D., 'The Jus ad Bellum and Russia's "Special Military Operation" in Ukraine,' *Journal of International Peacekeeping* 25 (2022): 121–127.

Golinkin, Lev, 'The Western Media Is Whitewashing the Azov Battalion,' *The Nation*, June 13, 2023, available at: www.thenation.com/article/world/azov-battalion-neo-nazi/

Gray, Colin S., *Another Bloody Century: Future Warfare* (London: Weidenfeld & Nicolson, 2005).

Gunaratna, Rohan, *Inside Al Qaeda: Global Network of Terror* (London: Hurst, 2003).

Hammes, Thomas X., *The Sling and the Stone: On War in the 21st Century* (St. Paul, MN: Zenith Press, 2004).

————, 'War Evolves into the Fourth Generation,' in Terriff et al., *Global Insurgency and the Future of Armed Conflict*, 21–44.

————, 'Fourth Generation Warfare Evolves, Fifth Emerges,' *Military Review* 87, no. 3 (2007): 14–23.

Hoffman, Frank, *Conflict in the 21st Century: The Rise of Hybrid Threats* (Arlington, VA: Potomac Institute for Policy Studies, 2007).

————, 'Combating Fourth Generation Warfare,' in: Terriff et al., *Global Insurgency and the Future of Armed Conflict*, 177–199.

————, 'Hybrid Warfare and Challenges,' *Joint Forces Quarterly* 52, no. 1 (2009): 34–39.

Howe, Herbert M., 'Private Security Forces and African Stability: The Case of Executive Outcomes,' *The Journal of Modern African Studies* 36, no. 2 (1998): 307–331.

Huntington, Samuel P., 'The Clash of Civilizations?' *Foreign Affairs* 72, no. 3 (1993): 22–49.

———, *The Clash of Civilizations and the Remaking of World Order* (New York: Simon & Schuster, 1998).

———, 'The Age of Muslim Wars,' *Newsweek*, 138, no. 25 (2001).

Kaldor, Mary, *New and Old Wars: Organized Violence in a Global Era* (Cambridge: Polity, 2001).

Kaplan, Robert, 'The Coming Anarchy,' *The Atlantic*, February 1994.

Kibbe, Jennifer, 'Conducting Shadow Wars,' *Journal of National Security Law & Policy* 5 (2012): 373–392.

Kilcullen, David, 'Countering Global Insurgency,' *Journal of Strategic Studies* 28, no. 4 (2005): 597–617.

Krepinevich, Andrew, *The Military Technical Revolution: A Preliminary Assessment* (Washington, DC: Center for Strategic and Budgetary Assessments, 2002).

Libiseller, Chiara, '"Hybrid Warfare" as an Academic Fashion,' *Journal of Strategic Studies* 46, no. 4 (2023): 858–880.

Lind, William S., 'Understanding Fourth Generation War,' *Military Review* (September/ October 2004): 12–16.

——— , 'Fourth Generation War Comes to a Theater near You,' *Chronicles Magazine* (October 2020), www.chroniclesmagazine.org/fourth-generation-war-comes-to-a-theater-near-you-1/

——— and Thiele, Gregory A., *4th Generation Warfare Handbook* (Kouvola, Finland: Castalia House, 2015).

Luna, Nathan, Vredenbregt, Leah, and Pereira, Ivan, 'What Is the Wagner Group? The "Brutal" Russian Military Unit in Ukraine,' *ABC News*, June 23, 2023, available at: https://abcnews.go.com/International/International/wagner-group-brutal-russian-military-group-fighting-ukraine/story?id=96665326

Luttwak, Edward, 'A Brief Note on "Fourth Generation Warfare,"' in Terriff et al., *Global Insurgency and the Future of Armed Conflict*, 52–53.

Mansoor, Peter, 'Introduction: Hybrid Warfare in History,' in: Williamson Murray and Peter R. Mansoor (eds.), *Hybrid Warfare: Fighting Complex Opponents from the Ancient World to the Present* (Cambridge: Cambridge University Press, 2012), 1–17.

Marten, Kimberly, 'Russia's Use of Semi-State Security Forces: The Case of the Wagner Group,' *Post-Soviet Affairs* 35, no. 3 (2019): 181–204.

McFate, Sean, *The Modern Mercenary: Private Armies and What They Mean for World Order* (Oxford: Oxford University Press, 2014).

———, *The New Rules of War: Victory in the Age of Durable Disorder* (New York: William Morrow, 2019).

Metz, Steven, *Rethinking Insurgency* (Carlisle, PA: U.S. Army War College Press, 2007).

——— and Kievit, James, *The Revolution in Military Affairs and Conflict Short of War* (Carlisle Barracks, PA: Security Studies Institute, 1994).

Mulford, Joshua P., 'Non-State Actors in the Russo-Ukrainian War,' *Connections* 15, no. 2 (2016): 89–107.

Münkler, Herfried, *The New Wars* (Cambridge: Polity, 2004).

Naftali, Timothy, *Blind Spot: The Secret History of American Counterterrorism* (New York: Basic Books, 2005).

Orenstein, Mitchell, *The Lands Between: Russia vs. the West and the New Politics of Hybrid War* (Oxford: Oxford University Press, 2019).

Organization of the African Union, 'OAU Convention for the Elimination of Mercenarism in Africa,' Libreville, Gabon, July 3, 1977.

Owens, Bill, *Lifting the Fog of War* (New York: Farrar Strauss Giroux, 2000).

Peters, Ralph, 'The New Warrior Class,' *Parameters* 24, no. 1 (1994): 16–26.

Rees, L.C., 'The End of the Rainbow: Implications of 5GW for a General Theory of War,' in: Daniel H. Abbott, *The Handbook of 5GW* (Ann Arbor, MI: Nimble Books, 2010), 20–29.

Rhagavan, Sudarsan, Morris, Loveday, Parker, Claire, and Stern, David L., 'Right-Wing Azov Battalion Emerges as a Controversial Defender of Ukraine,' *Washington Post*, April 6, 2022.

Samuels, Brett, 'Biden: Direct Conflict between NATO and Russia Would Be "World War III,"' *The Hill*, March 11, 2022, available at: https://thehill.com/policy/international/597842-biden-direct-conflict-between-nato-and-russia-would-be-world-war-iii/

Scahill, Jeremy, 'Bush's Rent-an-Army,' *Los Angeles Times*, January 25, 2007.

Schmitt, Carl, *Theory of the Partisan: Intermediate Commentary on the Concept of the Political* (New York: Telos Press, 2007).

Shawcross, William, *Deliver Us from Evil: Warlords and Peacekeepers in a World of Endless Conflict* (London: Bloomsbury, 2001).

Singer, Peter, *Corporate Warriors: The Rise of the Private Military Industry* (Ithaca, NY: Cornell University Press, 2003).

Smith, Eugene, 'The New Condottieri and US Policy: The Privatization of Conflict and Its Implications,' *Parameters* 32, no. 4 (2002): 104–119.

Taber, Robert, *War of the Flea: The Classic Study of Guerrilla Warfare* (Lincoln, NE: Potomac Books, 2002).

Thomson, Janice E., *Mercenaries, Pirates, and Sovereigns* (Princeton, NJ: Princeton University Press, 1994).

Thornton, Rod, 'Fourth Generation: A "New" Form of Warfare,' in Terriff et al., *Global Insurgency and the Future of Armed Conflict*, 87–94.

Turse, Nick, *The Changing Face of Empire: Special Ops, Drones, Spies, Proxy Fighters, Secret Bases, and Cyberwarfare* (Chicago, IL: Haymarket Books, 2012).

U.S. Department of Defense, 'Joint Vision 2010,' Washington, DC: U.S. Joint Chiefs of Staff (1996).

U.S. Department of Defense, 'Summary of the National Defense Strategy of the United States of America,' January 19, 2018, available at: https://dod.defense.gov/Portals/1/Documents/pubs/2018-National-Defense-Strategy-Summary.pdf

U.S. Senate, 'Report of the Senate Select Committee on Intelligence Committee Study of the Central Intelligence Agency's Detention and Interrogation Program,' December 9, 2014.

Van Creveld, Martin, *The Transformation of War* (New York: Free Press, 1991).

———, *The Rise and Decline of the Sate* (Cambridge University Press, 1999).

Vershenin, Alex, 'The Return of Industrial Warfare,' *RUSI*, June 17, 2022, available at: www.rusi.org/explore-our-research/publications/commentary/return-industrial-warfare

Williams, Jordan, 'Here's Every Weapon US Has Supplied to Ukraine with $13 Billion,' *The Hill*, August 26, 2022, available at: https://thehill.com/policy/defense/3597492-heres-every-weapon-us-has-supplied-to-ukraine-with-13-billion/

Wirtz, James J., 'Politics with Guns: A Response to T.X. Hammes' "War Evolves into the Fourth Generation,"' in Terriff et al., *Global Insurgency and the Future of Armed Conflict*, 47–51.

Zenko, Micah, 'Obama's Final Drone Strike Data,' *Council on Foreign Relations* (blog), January 20, 2017, available at: www.cfr.org/blog/obamas-final-drone-strike-data

2   Beyond the Fourth Generation

Fourth generation warfare has been defined as an evolved insurgency in which 4GW opponents focus on the political and moral dimension of a conflict to overcome the military superiority of regular armies.[1] The issue that predictably developed was the question of whether 4GW represented the final evolution of war or whether there could be further successive generations of warfare. Martin van Creveld's take on 4GW did not logically allow a further evolution of war since he postulated an end to the international system and a return to a more medieval world, but Hammes's theory did. At the end of his 2004 book, Hammes discussed the possibility that 4GW may be coming to an end and that it could be displaced by a new generation of warfare. He suggested: 'Fourth generation war has been around for more than 70 years; no doubt the fifth generation is evolving even as we attempt to deal with its predecessor.'[2] The chapter will explain some of the theories of nonviolent war formulated in the 1990s before outlining the reformulation of the generational warfare concept by Daniel Abbott and others in *The Handbook of 5GW*.

Theories of Nonviolent War

War is commonly defined as 'a prolonged state of violent conflict between two or more organized groups.'[3] This means that war has to be violent and it has to be carried out by and against larger political groups, such as states, nations, ethnic groups, or tribes. Clausewitz added the idea that war must be politically motivated and must be a means for achieving a political objective.[4] Van Creveld correctly criticized Clausewitz by pointing out that there are indeed nonpolitical wars, such as wars fought to establish justice, wars motivated by religion, and war for a community's existence.[5] Even the characteristic of violence as a requirement for war has been questioned in the context of cyber warfare, which usually does not result in violence against people.[6] There is no compelling reason as to why war would have to be always physically violent, far less that war would have to be conducted by specialists in violence, namely regular armed forces.[7]

Non-lethal Warfare

The idea of non-lethal warfare goes back some time. As early as the late 1940s, a U.S. Army study proposed the idea of psychochemical warfare, which would allow

DOI: 10.4324/9781003396963-3

an enemy population to be subdued by being drugged.[8] In the 1970s, law enforcement in the United States became interested in non-lethal weapons (NLWs) such as Tasers to subdue criminals without resorting to lethal force. In the early 1980s, the U.S. military started to research NLWs in view of new types of military challenges such as counterterrorism or counterinsurgency, where combatants need to be disabled without harming nearby civilians.[9] Some of the weapons that were developed include Tasers, rubber bullets, tear gas, acoustic weapons, and physical restraints. In the aftermath of the Battle of Mogadishu in 1993, the U.S. military created the Joint Non-Lethal Weapons Program, which coordinates the development of new types of NLWs.[10] The Pentagon now defines NLWs as 'weapons, devices, and munitions that are explicitly designed and primarily employed to incapacitate targeted personnel or materiel immediately, while minimizing fatalities, permanent injury to personnel, and undesired damage to property in the target area or environment.'[11]

The main argument for the military use of NLWs is the expanded spectrum of military missions that now include peacekeeping, counterterrorism, counterinsurgency, and military operations other than war (MOOTW), such as humanitarian assistance and disaster relief, all of which require the avoidance of the use of force or a minimum use of force. NLWs can enable armed forces or security forces to respond to limited threats in a less lethal manner, thereby protecting civilian bystanders and preventing the adversary from politically and propagandistically exploiting what could be seen as excessive use of force. NLWs could also make war more humane by substantially limiting bloodshed to situations where there is simply no alternative to lethal force.[12]

Robert Mandel argued that there are many motivations for governments to pursue war without bloodshed, such as casualty aversion; getting support from foreign governments, international organizations, and their own public for coercive action; better media coverage; and taking advantage of available technology.[13] There are several possible ways in which a bloodless war can be conducted: A war can be fought entirely remotely, by robots; it can be limited to cyberspace; or it can be conducted with weapons that incapacitate or psychologically coerce adversaries rather than kill them. Samuel Moyn argued that '[t]he American way of war is . . . defined by a near complete immunity from harm for one side and unprecedented care when it comes to killing people on the other.'[14] He also wondered whether the essence of war was 'control by domination and surveillance,' and whether future capability for the nonviolent coercion of an adversary would be an entirely good thing.[15] He asked: 'What if the elemental aim of endless war is not the death of the enemy soldiers but rather the potential nonviolent control of other peoples?'[16] If this was possible, would this not be war?

Neocortical Warfare

In 1997, the think tank RAND published a collection of articles that tried to capture the essence of warfare in the information age.[17] The ideas presented in the volume were far ahead of their time and are, therefore, worth revisiting. Among them is the concept of neocortical warfare, which was outlined by Richard Szafranski. He argued that 'military power resides in the domain of the mind and the will; the provinces of choice, "thinking," valuing or "attitude," and insight or "imagination" . . . military power can increase in effectiveness even as it decreases in violence.'[18] Clausewitz considered violence to be a means of compelling an enemy to do our will, which meant that greater levels of violence were assumed to have greater power to compel.[19] In other words, if the enemy suffered enough violence, then the enemy's will to resist could be broken.

Szafranski pointed out that violence is expensive, requires public support to use, and ultimately distracts from the real objective of war, which is 'subduing hostile will.'[20] Hence, it would make sense to focus on the problem of breaking the enemy's will rather than on devising better means for inflicting violence. What is important is to understand the enemy as a system or organism that has a will, which opens up the possibility of understanding and manipulating the enemy's will (or brain), making violence redundant. This approach is what Szafranski called neocortical warfare, defined as

> warfare that strives to control or shape the behavior of enemy organisms, but without destroying the organisms. It does this by influencing, even to the point of regulating the consciousness, perceptions and will of the adversary's leadership: the enemy's neocortical system. In simple ways, neocortical warfare attempts to penetrate adversaries' recurring cycles of 'observation, orientation, decision and action.'[21]

This makes it necessary to define war in terms of outcomes rather than the particular means that are used (armed forces). The point is that an enemy can be potentially subjugated without resort to military force or even the threat of military force.

Netwar and Noopolitik

The other notable concept in the RAND volume is called netwar and was developed by John Arquilla and David Ronfeldt. Netwar is conflict at the societal level waged by irregular, nonstate actors organized in networks, which provides them with an advantage over the more hierarchically organized nation states and their militaries. They defined netwar as

> an emerging mode of conflict (and crime) at societal levels, involving measures short of war, in which the protagonists use—indeed depend on using— network forms of organization, doctrine, strategy, and communication. These protagonists generally consist of dispersed, often small groups who agree to communicate, and act in a networked manner, often without a precise central leadership or headquarters. Decisionmakers may be deliberately decentralized and dispersed.[22]

Network organization can benefit greatly from the internet, but netwar is not simply war on the internet: It is a war of a network, conducted at the societal level, against a state or other opponent. The primary advantage of network organization is its great flexibility and adaptability. Networks can respond faster to opportunities and threats than hierarchical organizations, they can coordinate actions through swarming, they cannot be defeated through decapitation, and they can take offensive and defensive action, which may be observationally difficult to distinguish.[23]

Although Arquilla and Ronfeldt refer to terror and criminal organizations as capable of conducting netwar, they also point out that civil society actors such as NGOs and single-issue and activist groups can leverage netwar. A particular challenge for governments is to determine '[w]hose responsibility is it to respond? Whose roles and missions are at stake? Is it a military, police, intelligence, or political matter?'[24] Since the conflict plays out within a society and relies on nonviolent means to produce a political outcome, it is very hard to develop an appropriate strategy for a response that is nonviolent and that also minimizes the threat.

In 1999, Arquilla and Ronfeldt published a related short monograph titled *The Emergence of Noopolitik* that explored the concept of a noosphere and how it can be shaped in support of diplomacy and the conducting of conflict.[25] Noopolitik is, therefore, the politics of shaping the noosphere, which is a spiritual concept that can be described as the totality of human thought and ideas.[26] In other words, noopolitik is the application of geopolitics to the realm of information and ideas. More precisely, '[n]oopolitik is foreign-policy behavior for the information age that emphasizes the primacy of ideas, values, norms, laws, and ethics—it would work through "soft power" rather than "hard power."'[27] They argued that:

Noopolitik is an approach to statecraft, to be undertaken as much by nonstate as by state actors, that emphasizes the role of soft power in expressing ideas, values, norms, and ethics through all manner of media. This makes it distinct from realpolitik, which stresses the hard, material dimensions of power and treats states as the determinants of world order. Noopolitik has much in common with internationalism, but we would argue that the latter is a transitional paradigm that can be folded into noopolitik.[28]

To some extent, the emergence of a global noosphere can be considered to be a new phenomenon brought about by the invention of telecommunications and information technology, which connect all societies around the world and enable an unprecedented exchange of information.[29] In this new realm of a global information space, civil society actors would have an edge in terms of promoting ideas and advancing values. As a result, Arquilla and Ronfeldt suggested 'that noopolitik will require governments to learn to work with civil-society NGOs that are engaged in building cross-border networks and coalitions.'[30] The objective of noopolitics 'may ultimately be whose story wins.'[31]

Unrestricted Warfare

The term 'unrestricted warfare' was a Chinese military doctrine that originated in a book written by two Chinese military officers and published by the People's

Liberation Army (PLA) in 1999; it laid out China's strategy for defeating the United States by mostly non-military measures.[32] Robert Bunker, who reviewed the book, compared it to Mao Zedong's *On Guerrilla Warfare* and suggested that 'the significance of this work cannot be overstated.'[33] Robert Spalding wrote an entire book of commentary on Qiao Liang and Wang Xiangsui's book, *Unrestricted Warfare*, suggesting that '[o]ne can find glimpses into the secretive mentality of the CCP leaders, but this one is the single most important book for understanding the China of today.'[34] Donald Reed argued that 5GW 'closely models the description by Qiao and Wang of unrestricted warfare.'[35]

Ruleless War

Many China analysts, most notably Michael Pillsbury and Robert Spalding, have concluded that China seeks to supplant the United States as the global hegemon by the middle of the 21st century.[36] The Chinese Communist Party (CCP) understands that it is unlikely to win a direct military conflict with the United States, which means that it has to weaken the United States to the point that an effective resistance to CCP objectives becomes impossible. In their preface, Qiao and Wang claim:

> [If] we acknowledge that the new principles of war are no longer 'using armed force to compel the enemy to submit to one's will,' but rather are 'using all means, including armed force or non-armed force, military and non-military, and lethal and non-lethal means to compel the enemy to accept one's defeat.'[37]

The primary argument of their book is that conventional warfare has become obsolete, and that in the new mode of warfare there would be no rules. Qiao stated in an interview that 'the first rule of unrestricted warfare is that there are no rules, with nothing forbidden.'[38] As a result, the distinction between war and peace has not just become blurred, it is meaningless. The CCP considers itself to be in a constant war with the United States, which can only end with the CCP either achieving global domination or facing defeat. Spalding suggested, 'China is fighting us on every front, in ways visible and invisible.'[39]

Omnidirectional Attack

A key assumption is that 'the metamorphosis of warfare will have a more complex backdrop' owing to the 'reduced use of military force to resolve conflicts.'[40] Qiao and Wang claim '[t]he era of "strong and brave soldiers who are heroic defenders of the nation" has passed.'[41] The new fighters will be 'non-professional warriors' such as computer hackers, The book argues that a number of non-military actions such as cyber warfare, stock market manipulation, and media reporting of conflicts could constitute types of future warfare.[42] Future war is going to be so multidimensional (physical and virtual dimensions) and multimodal (military and non-military modes of fighting) that it would become impossible to label it in any meaningful way. The authors argue that high-tech weaponry and high lethality would not be particularly useful in future wars, and they instead point at the advantages of new concept weapons (electromagnetic, computer, and media weapons) and non-lethality.[43]

The battlefield would be everywhere, and wars would be fought by non-professional or non-military warriors, who could be hackers, terrorists, or financiers. Modern societies are uniquely vulnerable to innovative modes of indirect attack, all of which are designed to weaken an adversary in the long term to the point that a potential subsequent military attack (should one be necessary) could succeed. Below are some of the modes of attack discussed by Qiao and Wang:

- *Trade war*:

  [T]he use of domestic trade law on the international stage; the arbitrary creation and dismantling of tariff barriers; the use of hastily written trade sanctions; the imposition of embargoes on exports of critical technologies; the use of Special Section 301 law; and the application of most-favored nation (MFN) treatment, etc.[44]

  The authors suggest that the United States has perfected this type of warfare, which is meant to undermine adversarial economies and their respective technological and military capabilities.
- *Financial war*: '[F]inancial war is a form of non-military warfare which is just as terribly destructive as a bloody war but in which no blood is actually shed.'[45] Financial war is about attacking foreign financial markets and currencies in order to induce financial and social chaos and inflict financial and economic damage. The Asian financial crisis of 1998 is cited as an example of financial war wreaking havoc upon several industrialized Asian countries. Today, a key U.S. concern is the attack on the U.S. dollar by Brazil, Russia, India, China, and South Africa (the BRICS nations), which are selling off United States-denominated assets and increasingly circumvent the U.S. dollar for international payments.[46]
- *Terrorism*: 'The advent of bin Laden-style terrorism has deepened the impression that a national force, no matter how powerful, will find it difficult to gain the upper hand in a game that has no rules.'[47] Qiao and Wang point out that 'new terror war' may use new technological 'superweapons,' including chemical weapons, biological weapons, and computer hacking and information warfare.
- *Ecological war*:

  Ecological war refers to a new type of non-military warfare in which modern technology is employed to influence the natural state of rivers, oceans, the crust of the earth, the polar ice sheets, the air circulating in the atmosphere, and the ozone layer.[48]

  Qiao and Wang suggest that environmental modification technologies exist or could be developed to induce ecological disasters as a means of covertly attacking adversaries with earthquakes, flooding, droughts, or hurricanes.

- *Psychological warfare*: 'Even the last refuge of the human race—the inner world of the heart—cannot avoid the attacks of psychological warfare.'[49] The method could involve 'spreading rumors to intimidate the enemy and break down his will.'[50] Qiao and Wang acknowledge that psychological warfare is old but state that there are novel ways of using it, as was demonstrated by the U.S. military in the Gulf War.
- *Smuggling warfare*: This includes 'throwing markets into confusion and attacking economic order.'[51] A foreign economy could be flooded with pirated or illicit goods, which reduces the profitability of foreign companies and distorts markets. A variation of this could be the smuggling of human beings to distort labor markets and undermine the political stability of a country.
- *Media warfare*: This involves 'manipulating what people see and hear to lead public opinion along.'[52] This can be accomplished by sponsoring foreign-language media targeted at an adversary's society, co-opting foreign media through financial leverage/advertising revenues, and cultivating or intimidating foreign journalists to report China's preferred narratives.
- *Drug warfare*: This means 'obtaining sudden and huge illicit profits by spreading disaster in other countries.'[53] Qiao and Wang pointed out that the British launched a war against the Qing dynasty to protect 'national drug trafficking activity on probably the grandest scale in recorded history.'[54] The current explosion of drug-overdose deaths (over 100,000 per year in the United States as of 2022) resulting from fentanyl that originates in China must be considered in the context of drug warfare. According to Robert Spalding, '[d]rug traffickers use two techniques for delivering fentanyl manufactured in China: It is either shipped directly into the United States via international mail or shipped into Mexico to be smuggled into America.'[55]
- *Network warfare*: This involves 'venturing out in secret and concealing one's identity in a type of warfare that is virtually impossible to guard against.'[56] What is referred to here is cyber warfare, which seeks to subvert or attack foreign computer networks for the purpose of espionage and sabotage.
- *Technological warfare*: This concerns 'creating monopolies by setting standards independently.'[57] Spalding claims that 'China is trying to dominate international standards for the 5G wireless networks' in order to control a key component of a nation's telecommunications infrastructure. This could enable China to access all kinds of data derived from the Internet of Things and even enable it to exercise control over devices connected to 5G networks in order to sabotage them.[58]
- *Fabrication warfare*: This means 'presenting a counterfeit appearance of real strength before the eyes of the enemy.'[59] The concept is not explained any further in *Unrestricted Warfare*, but it clearly relates to military deception and media warfare with the objective of intimidating an adversary through fabricated strength.
- *Resources warfare*: This is 'grabbing riches by plundering stores of resources.'[60] According to Spalding, 'China has been trying to dominate the market for increasingly essential rare earth.'[61] In particular, China is trying to gain control over the world's lithium resources since lithium is essential to battery technology and, by extension, electric vehicles.

- *Economic aid warfare*: This involves 'bestowing favor in the open and contriving to control matters in secret.'[62] China has been expanding its economic ties with African nations with a view to securing access to raw materials by financing, constructing, and controlling infrastructure through the Belt and Road Initiative. It is an 'economic hitman' approach in which China uses debt incurred by developing nations as leverage to force political and social changes, as well as to eventually take over the infrastructure if debts cannot be repaid.[63]
- *Cultural warfare*: This includes 'leading cultural trends along in order to assimilate those with different views.'[64] China has been promoting its own style of authoritarianism in other countries and is exerting cultural influence through its Confucius Institutes, which can be found on many Western university campuses.[65]
- *International law warfare*: This involves 'seizing the earliest opportunity to set up regulations.'[66] Chinese legal warfare has become the subject of extensive academic discussion, in particular in relation to Chinese abuse of international law to assert territorial claims in the South China Sea and restrict U.S. naval operations in proximity to the Taiwan Straits and the artificial islands claimed by China.[67]

Supra-combinations

Unrestricted warfare is based on the combination of different actors, domains, means, and levels of conflict. Combination makes it possible to create synergies and achieve far greater effects than can be done without combination. In other words, the combination is more than just the sum total of its constituents.

- *Supranational combinations*: 'This method, resolving conflicts or conducting warfare not just with national power, but also with combinations of supranational, trans-national, and non-state power, is what we mean by the general term supra-national combinations.'[68] This is the art of assembling alliances or coalitions comprising different nations, international organizations, multinational corporations, NGOs, and other nonstate actors, such as terror organizations, criminal syndicates, and hacker groups.[69] Again, Qiao and Wang's main example is the Gulf War, this time in view of the United States' political and perceptional maneuvering to secure international support for an attack on Iraq.
- *Supra-domain combinations*: 'From this point of view of beyond-limits thinking, "supra-domain combinations" means the combining of battlefields. Each domain may, like the military domain, constitute the principal domain of future warfare.'[70] The domains mentioned include 'politics, economics, the military, culture, diplomacy, and religion.'[71] In short, war requires action across many domains (beyond the traditional military domains) in order to exert maximum pressure on an adversary.
- *Supra-means combinations*: '[I]n a world of unprecedented complexity, the form and the scope of application of means is also in a state of continuous change, and a better means used alone will have no advantage over several

means used in combination.'[72] Qiao and Wang point out that, when choosing a means combination, it only matters whether it can achieve a desired objective, and that one cannot always rely on ready-made means such as economic block-ades. One has to be creative in the creation and combination of means.

- *Supra-tier combinations*: Qiao and Wang argue that war can be divided into four distinctive levels: Grand war—war policy (grand strategy); war—strategy; campaigns—operational art; and battles—tactics. 'Our issue is how to use some method to break down all the stages [levels of war], and link up and assemble these stages at will.'[73] The idea is to act across the different levels of war, which means that actions at the tactical level may have strategic or grand strategic impacts.

Unrestricted Warfare and 5GW

The unrestricted warfare concept developed some key ideas that are highly relevant for 5GW, but it also differs in many ways from it. The fundamental idea that un-restricted warfare and 5GW have in common is what Basil Liddell-Hart called the 'indirect approach,' or the idea that the best way of achieving victory is by avoid-ing a direct confrontation with an enemy's main forces and focusing on attacking logistics and morale.[74] Both theories, unrestricted warfare and 5GW, advocate for supra-combinations to achieve a disproportionate effect on an adversary. Finally, both argue in favor of the limited use of violence, not for moral reasons but for pragmatic reasons, namely the economy of force and the need to control escalation. Where 5GW departs from unrestricted warfare is in terms of the focus on the hu-man domain to the neglect of other domains of war. Economic or financial means could be part of a 5GW strategy, but the objective would not be economic or finan-cial: The objective would be psychological and cultural more than anything else. The main battlespace of 5GW is the human mind, which is the realm of perception, cognition, culture, and social relationships.

Towards a Theory of 5GW

The beginning of the War on Terror led to a brief revival of the 4GW school owing to Western interventions in developing countries, with counterinsurgency opera-tions against globally networked terror and insurgent groups. However, the 9/11 and Amerithrax attacks of 2001 also appeared to indicate another transformation in warfare and insurgency rather than representing the kind of evolved Maoist in-surgency of 4GW. These terror attacks involved nonstate actors attacking a power-ful state actor by way of weaponizing civilian technology or by exploiting stolen nation-state weapons capabilities.[75]

Hammes stated that 'super-empowered individuals or small groups would be in keeping with several emerging global trends—the rise of biotechnology, the increased power of knowledge workers, and the changing of loyalties.'[76] Hence, earlier discussions of a potential fifth generation of warfare focused on the so-called 'new terrorism,' which was supposed to be more deadly, more resilient, and harder to fight than the old terrorism of politically or ideologically motivated revolutionary/separatist groups because of globalization, easier access to advanced technology, and religious motivations.[77] Donald Reed made the first attempt at developing a coherent theory of 5GW. He argued that

> fifth generation warfare allows the battlefield to become omnipresent by expanding it beyond the domains of third and fourth generation warfare, it also empowers any entity with the economic and technical means, and motivated by self-interest, with the ability to wage war. War is no longer the monopoly of nation-states, or even of insurgencies, but as a result of the Information Age and the effects of globalization has become the province of both state and non-state entities in the form of networks, super-empowered individuals, and groups that are capable of forming supra-combinations that have not been seen before.[78]

Some of the key ideas in Reed's article are that the role of militaries in 5GW is greatly diminished, that 5GW blurs all boundaries, and that the inferior can defeat the superior.[79] What Reed did in his 2008 article was largely a reformulation of *Unrestricted Warfare*. In 2010, Daniel Abbott moved the debate forward by publishing a collection of theoretical essays that tried to locate 5GW properly within the 4GW framework while also revising the earlier ideas of generational war.[80]

The OODA Loop and Fifth Generation Warfare

The fighter pilot and strategist John Boyd developed the observe–orient–decide–act (OODA) loop as a model for decision-making, based on his experience in air-to-air combat and developed in the 1980s and 1990s (Boyd died in 1997).[81] Many 4GW and 5GW proponents, including Lind and Abbott, have invoked the OODA loop in order to highlight differences in the generations of warfare.

The key idea is that there are distinctive steps that have to be completed in decision-making, forming a decision cycle. The primary input comes from the surrounding world or operational environment. *Observe* means that the decision-maker has to monitor the environment or collect data on the environment. *Orient* means that analysis has to be performed to make sense of the sensory or external data to provide a perspective on the specific situation. *Decide* means to choose from a number of possible courses of action based on the orientation or analysis step. *Act* means to implement the decision. Lind suggested that

> whoever can orient more quickly to a rapidly changing situation acquires a decisive advantage because his slower opponent's actions are too late and therefore irrelevant—as he desperately seeks convergence, he gets ever increasing divergence. At some point, he realizes he can do nothing that works. That usually leads him either to panic or to give up, often while still physically largely intact.[82]

Act   Observe

Decide   Orient

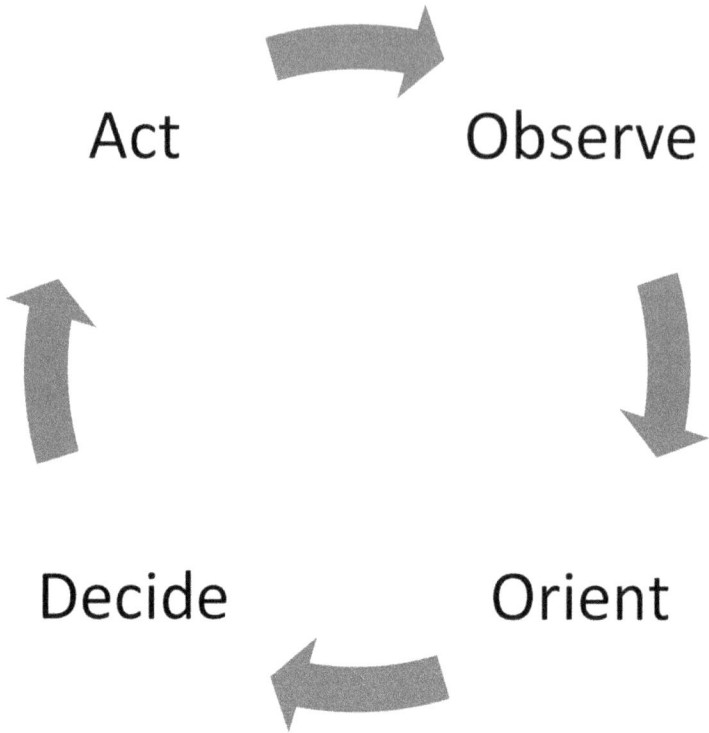

Figure 2.1 OODA Loop
Source: Created by author

Different generations of warfare target different steps in the OODA loop in order to disrupt or slow down the adversary's ability to complete the decision cycle.[83] Abbott suggested that '[w]ar is going deeper into enemy minds. Every generation of warfare aims for deeper in the enemy's OODA loop.'[84] First generation warfare targeted the acting phase, 2GW targeted the decision phase, 3GW targeted the orientation phase, and 4GW also targeted the orientation phase but additionally targeted observation.[85] According to Chad Kohalyk, '5GW will attack an enemy's ability to OBSERVE. The enemy could be blind, unaware to the true identity of the adversary he is engaging with, or maybe oblivious to the fact that he is fighting a war at all.'[86] In other words, 5GW targets perception and therefore undermines the ability of adversaries to make good decisions (or any relevant decisions at all).

The xGW Framework

The main problem with launching a theory of 5GW was that the whole concept of generational and post-Westphalian war had been tarnished by the justified criticism of notable military historians and scholars. What was needed was a theory less tied to contentious interpretations of history and more abstract. *The Handbook of 5GW* therefore suggested dropping the term 'generation' altogether and replacing it with gradient or grade, which means '[a] stage or degree in a process.'[87] A gradient is defined as '[a] stage or degree in a process'; '[a] position in a scale of size, quality or intensity'; '[a]n accepted level or standard.'[88] Warfare moves between a range of gradients. Each change in gradient requires 'a complete societal change.'[89] The first gradient requires a division of labor, the second industrialization, and the third nationalism.[90] The fourth gradient seems to require network organization, and the fifth globalization.

Warfare can move upwards to a higher gradient and it can also move downwards to a lower gradient. Furthermore, different gradients or generations of warfare may temporarily coexist: The emergence or appearance of a higher gradient of warfare does not completely displace the previously dominant gradient of warfare.[91] This means that the history of warfare should not be considered a sequential evolution of different modes of war, which was a major criticism of the 4GW school.[92] According to Abbott,

> The xGW framework rejects the theory of sequential emergence, and the Generations of Modern War (GMW) school that is associated with it. While some theorists in the xGW framework still use the term generation, the elements of the taxonomy are now generally known as gradients. The gradients of war, like gradients we see in other elements of social organization (wealth, height, skin color, and so on) flow indefinitely into each other, and their emergence pre-dates written history.[93]

The new starting point then becomes zeroth gradient war (0GW), which is defined as total war or genocide, followed by first gradient war (1GW), which is a 'concentration of labor,' second gradient war (2GW), which is a 'concentration of firepower,' third gradient war (3GW), which relies for victory on 'better minds' (ideas and information), and fourth gradient war (4GW), which blurs war and peace.[94] 'In 5GW violence is so dispersed that the losing side may never realize that it has been conquered.'[95]

In other words, higher gradients may exist in earlier historical periods and may also coexist with modes of warfare of a lower gradient. Warfare may also regress from a higher gradient to a lower gradient or advance to an even higher gradient. When the potential of 5GW is exhausted, a conflict may degrade to 4GW or 3GW or it may advance to 6GW. Key to the differentiation of the gradients is the idea of complexity: Higher-gradient warfare is more complex than lower-gradient warfare, which also affects the amount of violence necessary to achieve objectives: 'In the xGW framework, kinetic force has greater utility at lower gradients of the framework and less utility at the higher gradients of the framework.'[96] Still, xGW thinkers 'agree that 5GW will be a form of conflict that would have been impossible in the past' owing to the role of advanced technology.[97] A caveat here is that 5GW is not closely tied to any particular technology such as the internet, although the internet can enable new modes of conflict (cyber warfare) that fit in the 5GW paradigm and, hence, were not possible earlier.

Another advantage of the xGW framework is that it can accommodate indefinite future modes of war and can expand and adjust the theory in light of new developments. Sixth generation warfare will be able to overcome 5GW and so on. Even some defining characteristics of 7GW and 8GW could be deduced from the internal logic of the framework: The next generation must be able to counter or render obsolete the previous one; it also must be more complex, and violence must be even more dispersed, precise, or indirect than in the previous one. This leads to speculation that 6GW could be cyber warfare and electronic warfare directed against critical infrastructure, and 7GW could be conflict fought by AI systems, with limited or no human participation. Of course, it is very difficult to predict future gradients of warfare that have yet to materialize.

Secrecy in 5GW

Unlike 4GW, which is a political insurgency and as such strives for visibility—there is no question as to the existence of conflict or to the objectives of the 4GW opponents—the matter is completely different with respect to 5GW opponents, who may be motivated by some political ideology or a belief system, but who may have no interest in publicizing it or the fact that they would be engaged in a war. 'There are no warnings, no communiques, no explanations—or at least none that can be trusted—only events which may or may not be random.'[98] Fifth generation warfare opponents prefer to remain hidden and, if that is not possible, they prefer to frame their aggressive actions as being ultimately benign. Ideally, the victims of a 5GW mode of attack would never be aware that they were under attack, or, if they were aware, they would never figure out from where the attack was actually coming and who was responsible. L.C. Rees suggested that 5GW is a 'war [in which] the enemy may fight you with a form of war that you not only can't see, but even worse, don't even believe in,' resulting in the following: 'You never knew you were at war. You never saw what hit you. You never knew there was a chance for victory. You never knew that you were defeated. You don't believe in any of the above.'[99]

Fifth generation warfare seems closely related to the practices of statecraft described as covert action, covert operation, or active measures, in the sense that sponsors of these activities seek some degree of plausible deniability, either by making it very difficult to trace activities back to them or by relying on the ambiguity inherent in the activities. A lot can be learned from studying CIA covert action or Soviet active measures in terms of how a connection to sponsors, their objectives, and some of the activities themselves can be hidden. For example, in the realm of covert operations, it is standard practice to rely on front organizations that are specifically created to provide cover for personnel and activities, to recruit agents of influence who can influence perceptions and decision-making, and to rely on the use of information as a weapon (propaganda or disinformation). Rory Cormac and Richard Aldrich have pointed out that covert action does not need to rely on complete secrecy, as in hiding the sponsorship of activities from everybody

Table 2.1 The Seven Gradients of War

Gradient of War	Domains	Adversaries	Objectives	Violence
Zeroth	Land	Tribes/populations	Total destruction of enemy	Genocide
First	Land/sea	Armies/navies	Direct destruction of close enemy forces	Kinetic force against military forces
Second	Land/sea	Armies/navies	Indirect destruction of enemy forces by firepower	Kinetic force against military forces
Third	Land/sea/air/cyber	Combined armies/ militaries	Destruction of the enemy military command and control	Precision targeting of military command and control
Fourth	Political	Networks	Destruction of the enemy's political will to fight	Kinetic and non-kinetic means
Fifth	Physical/information/ cognitive/social	Supra-combinations	Destroy or render the enemy's efforts irrelevant by any means	Mostly non-kinetic, with some dispersed violence
Sixth	Technical systems	Deep states/state/non-state combinations	Co-opt, take control of, or destroy the enemy's technical systems	Non-kinetic and non-contact

Source: Adapted from Donald Reed, 'Beyond the War on Terror: Into the Fifth Generation,' *Studies in Conflict and Terrorism* 31, no. 8 (2008): 691.

other than those who have been 'cleared' to know.[100] To some extent, the exposure of covert activities is always expected to occur at some point.[101] Exposure of covert activities (not their official acknowledgment) might not diminish their effectiveness as it creates yet more uncertainty.[102] Many people will remain unaware that an exposure occurred, will even dismiss information related to exposure, or will simply deny that any harm was done.[103]

Many aspects of 5GW activities will not be secret as such—they will be observable by many audiences. The main challenge is the intellectual fight over interpretation of these activities—namely, their purpose and their effects, which a 5GW opponent will do everything they can to paint as benign and, if negative effects become visible, to declare them to be minor or unintended. The ambiguity that the approach creates makes it hard for a victim or analyst to comprehend that there is a war, and that intentional harm has occurred and was significant enough to produce a desired outcome for an adversary.

This ambiguity—rather than outright secrecy—inherent in 5GW creates an epistemological problem: How can one prove the existence of 5GW? It makes sense to think about 5GW in terms of effects and outcomes, as these are usually observable while sponsors, intentions, and particular means may not be observable. The outcomes can provide clues with respect to intent, and intent can potentially reveal the real actors. Abbott suggests that '[q]uantitative research can be used to study 5GW.'[104] In particular, '[q]uantitative data analysis could range from simple procedures, such as multiple regression and analysis of variance (ANOVA), to more complex procedures such as multivariate analysis and structural equation modeling.'[105] Data points relevant to understanding of the effects of a 5GW attack could be opinion surveys; various demographic data, including unexplained changes in life expectancy, excess mortality, or fertility; and also crime statistics. One could also consider the probability of an unlikely type of negative event or accident occurring repeatedly, as well as the unwillingness or inability of authorities to investigate, explain, mitigate, or address the issue. If bad things keep happening for no apparent reason, it is likely intentional. If no action is taken to deal with a threat, complicity of those tasked to prevent threats is likely.

Command and Control in 5GW

Fifth generation warfare proponents generally agree that it is a war fought by networks across different actor groups, different domains, and different levels of war and using different means, and 'that a 5GW force becomes increasingly effective the more disparate its efforts become.'[106] This raises the question as to how effective coordination can be achieved in a way other than through some centralized command structure, which could not possibly be able to coordinate every aspect of a multimodal and omnidirectional attack owing to the immense complexity involved. The answer can be found in the theory of complex systems, swarming behavior, and chaos theory. As Shane Deichman argued, a 5GW opponent may rely on the 'naturally emerging behavior of complex systems' to synchronize efforts across different networks of people and organizations. In other words, it is generally wrong, in 5GW, to assume the existence of some grand conspiracy or some evil mastermind pulling all the strings in the background. The truth is much more mundane: Diverse sets of autonomous actors can be co-opted through overlapping interests, resulting in some degree of self-synchronization or swarming behavior.[107] Autonomous actors simply watch what other actors within their network do in order to adjust their own behavior accordingly. Nobody in the swarm needs to have a complete plan for coordinating their actions. Everybody in the swarm just follows simple rules: Move in the same direction as others, remain close to neighbors, and avoid collisions.[108] In a netwar, this is mostly done by public signaling through publications, public statements, or public actions.

Co-optation may occur through the use of financial incentives (bribes or opportunities to profit), through a shared ideology or compatible political objectives, or through claiming a shared or friendly identity (racial, ethnic, or religious).[109] Many organizations or individuals that have been co-opted by a 5GW opponent may not even be aware that they serve someone else's interests since they may genuinely believe that their actions are either purely self-interested or serve a purpose other than the real intent of the 5GW opponent. According to David Axe, 5GW

> is less clearly ideological but just as sweeping in its goals. 5GW is when a party exploits or encourages an existing or emerging crisis to achieve strategic goals that those most directly involved in the crisis might not even be aware of. 5GW is a form of stealthy proxy war.[110]

One can take an existing group and co-opt it by offering incentives or support, and/or one can exacerbate existing problems to move things along in a direction in which they were already going. The advantage of intervention is that the intervener has an opportunity to influence the outcome, while just allowing events to unfold by themselves leaves too much to chance.

A good example of how self-generation and synchronization across different organizations and groups can happen spontaneously is the way in which the East German secret police (commonly referred to as 'Stasi') carried out a 'systematic and persistent subversion of the West European [peace] movement' from 1979 to 1990.[111] The general idea was to use European peace activists to put pressure on West European governments to prevent the deployment, and later call for the removal, of the American intermediate range nuclear weapons from European soil, following NATO's 1979 'double-track' decision. The Soviet Union created urgency and an emergency by emphasizing how greatly American missile deployments threatened international stability and increased the risk of global thermonuclear war, which made it easier to co-opt existing peace activist groups, who were genuinely concerned about nuclear war and who were not necessarily communists. Key to the successful creation of a strong anti-nuclear weapons movement was the 'Generals for Peace' initiative, which was launched by a former West German naval officer and historian, Gerhard Kade, who was also a Stasi and KGB agent. He managed to assemble an impressive collection of over a dozen unwitting former NATO generals and admirals, who publicly warned of the dangers of nuclear war and who advocated for nuclear disarmament. This resulted in a large amount of publicity activity, with conferences, media debates, books, and other publications, creating a much larger movement that the Stasi did not control.[112] In fact, the Stasi became concerned that the West European peace movement could spill over to East Germany and cause problems.[113]

The New 5GW Belligerents

Discussions of hybrid warfare, gray zone conflict, or covert action always assume that a state actor is ultimately responsible by controlling or empowering a set of state or nonstate proxies. In 5GW, it is likely that states are participants in conflict or that certain state-specific capabilities are used, but it does not mean that states are the ultimate initiators, controllers, or sponsors. Fifth generation warfare is group-on-group warfare, and this can mean that states or certain parts of a state's bureaucratic apparatus are co-opted to act on behalf of one group against another group. This can be an inter-societal conflict (a culture war) or an inter-civilizational conflict.[114] States therefore tend to be the victims of 5GW rather than the genuine perpetrators of 5GW. As Daniel McIntosh pointed out, in 5GW '[t]he conflict is not to conquer the state, or to divide the state, but to undermine the state.'[115]

Super-empowered Individuals

It is consistent with the idea that violence is more dispersed in higher-gradient warfare that individuals or small groups would be dominant actors in 5GW. Hammes suggested that super-empowered individuals could be 'one possible form of future war' or a major characteristic of 5GW.[116] The term 'super-empowered individual' was coined by Thomas L. Friedman in his 1999 book *The Lexus and the Olive Tree*, which explored globalization. He wrote that

> [t]he greatest danger that the United States faces today is from Super-Empowered individuals who hate America more than ever . . . today's Super Empowered Angry Man or Angry Woman can use the powers embedded in globalization to attack even a superpower.[117]

Super-empowered individuals are potentially dangerous because of the globalization and technology that give them global reach (cheap transportation and communication) and the kind of destructive power only nation states could previously leverage. Super-empowered individuals are also a very diverse group of actors in international relations, ranging from drug lords such as Pablo Escobar to terrorist leaders such as Osama bin Laden, transparency activists such as Julian Assange, and major financiers and industrialists such as George Soros, Bill Gates, and Jeff Bezos.[118]

What they have in common are their lack of formal or political power and their ability to shape the world, influence world events, leverage technology, and act globally. They have important resources at their disposal in terms of unique expertise, connections and access to other powerful individuals and information, perhaps great financial wealth, and access to inventions and powerful technology. According to a U.S. Army TRADOC report from 2017, super-empowered individuals have the following characteristics:

• They are '[h]ighly connected and able to reach far beyond their geographic location.'
• They have '[a]ccess to powerful, low-cost commercial technology.'
• They are '[o]ften more difficult to trace or attribute responsibility for actions.'
• They are '[n]ot beholden to nation-state policies, ethics, or international law.'
• They have '[v]arying motivations (political, ideological, economic, and monetary).'
• They are '[o]ften unpredictable, may not operate or execute like a traditional rational actor.'[119]

Particular technologies that originated from, or are being developed in, the private sector and that are very difficult for governments to monitor, regulate, and control can empower individuals and small groups in a completely unprecedented manner, which includes nanotechnology, biotechnology, information technology, and cognitive (neuro) technologies.[120]

Criminal Organizations

Fifth generation warfare transcends many traditional boundaries such as between war and peace, military and civilian, covert and overt, and the licit and illicit. Criminal organizations can engage and have engaged in 5GW against states.[121] Even though criminal organizations generally lack political objectives, they are often important in 5GW because of their unique set of capabilities: They can corrupt officials, they control smuggling routes, they have access to parts of the legitimate economy through control of legal businesses, and they can operate globally and clandestinely. Criminal organizations also tend to use violence strategically and selectively—they understand that violence is costly, may invite retaliation, and has only short-term benefits.

Criminal syndicates can be found in most countries of the world. They are internationally connected; they specialize in the control of illicit markets such as illegal narcotics, human trafficking, prostitution, and gambling; and their activities are generally transnational in orientation. According to the U.S. Department of Justice,

> The most powerful international organized crime groups benefit from the symbiotic relationship that their leaders have developed with corrupt public officials and business tycoons. The three elements combine forces to form strategic alliances. International organized crime in its highest form is far removed from the streets. These groups are highly sophisticated, have billions of dollars at their disposal, are highly educated, and employ some of the world's best accountants, lawyers, bankers and lobbyists. They go to great lengths to portray themselves as legitimate businessmen and even advocates/benefactors for the local populace and others.[122]

It is important to acknowledge that organized crime cannot exist without at least some tacit approval from governments, which means that these organizations can sometimes reach so deep into governments that it can become difficult to determine where a criminal enterprise ends and a legitimate government begins.

Leading criminals can even be members of a government, or senior government officials can be in the pockets of criminal organizations. For example, in the early 1980s, Colombian drug lord Pablo Escobar was not just one of the world's richest people, he even became a member of the Colombian Congress in 1983.[123] He also constructed a neighborhood for the poor in Medellín, which gained him much popularity. In neighboring Bolivia, drug traffickers sponsored a coup to install a co-operative government under General Luis García Meza in 1980, and in Chechnya the corrupt government of Dzhokhar Dudaev created a safe haven for criminal gangs from 1991 to 1996 in order to profit from crime.[124]

Even a major country such as Russia has been accused of being a criminal state owing to the apparent penetration of the government by criminal groups and criminal interests in the 1990s. In the Yeltsin period, organized crime was extremely pervasive in Russia, with some 8,000 criminal formations producing so much revenue that it represented a quarter of Russia's GDP. These groups controlled both licit and illicit businesses, including in important industry sectors such as oil and gas.[125] According to Yuriy Voronin,

> [i]n the highest echelons of the Russian government, crime and politics are often indistinguishable from one another. Crime in the post-Soviet era, in other words, is often a continuation of politics by other means—but carried out by the same persons: the functionaries, security and police officials and other who run the state. Professional criminals commonly manipulate police and customs organizations, military establishments, and financial institutions. Furthermore, the tentacles of organized crime apparently extend to the highest levels of government.[126]

Although the Russia of the 1990s was an extreme case of rampant criminality, there is no doubt that criminal organizations can corrupt government officials everywhere. What makes organized crime so dangerous is that its influence is often well hidden behind a façade of respectability. Once the façade breaks down owing to exposure of corruption or excessive violence that goes unpunished, the result is a steady corrosion of the legitimacy of governments. Ironically, 'transnational organized crime can prosper under political and economic unrest because conflict zones exhibit loosened controls, sizable capital availability due the need for hard cash, and significant returns on investment due to high risks.'[127] In short, organized crime can benefit from the destruction of states, which makes it a perfect ally for 5GW opponents.

Deep States

The notion of a 'deep state' or 'independent security state' has become both prominent and controversial in recent years, resulting also in some confusion as to what it actually means.[128] The simplest way of understanding the deep state is to refer to the older and more established concept of the 'military–industrial complex,' which is a term that was first coined by President Eisenhower in his January 1961 farewell speech.[129] Eisenhower famously stated: 'In the councils of government, we must guard against the acquisition of unwarranted influence, whether sought or unsought, by the military–industrial complex. The potential for the disastrous rise of misplaced power exists and will persist.'[130] The deep state is, therefore, not a conspiracy or a conspiracy theory, but rather it is an institutional arrangement that is based on diminished democratic accountability, which has enabled bureaucratic actors to achieve some degree of autonomy from the rest of the government.[131] Deep states exist in many countries and they differ in their power, influence, and characteristics, but they are always unified and somewhat separate from the rest of the visible or constitutional government.

According to Sean McFate, '[d]eep states are like states with cancer. Their institutions of power—the military, the judiciary, intelligence agencies—have gone rogue; rather than serving the state they make the state serve them.'[132] The deep state is hence a convergence of bureaucratic and corporate interests that is sometimes antithetical to the best interests of the citizenry owing to threat exaggeration that drives a constant expansion of national security spending, a constant expansion of the surveillance state, and a constant pursuit of conflicts that justify the former. Deep states are most powerful in a state of emergency as the normal rules of government accountability are suspended. According to Carl Schmitt's logic that 'sovereign is he who decides on the state of exception,' the deep state is the real sovereign.[133]

Michael Glennon described the American deep state as a 'Trumanite network' because it was originally formed during Truman's presidency.[134] It is essentially a network of individuals who are well placed in the national security apparatus, holding key positions in the mid and upper management, who share a similar background and belong to the same 'tribe,' who practice a cult of secrecy, and who ultimately protect their own interests against outsiders, both within and outside the government. Deep states usually have a militarized culture in which conformism and obedience thrive. This means that they tend to protect the status quo and resist change.[135]

Glennon claimed that there is now a dichotomy, with a Madisonian government comprising the public institutions that are defined by the constitution and then a secret government composed of the Trumanite network that controls national security policy, which leads to a complex interplay between the two. 'For the double government to work, the Madisonian institutions must seem in charge, for the Trumanites' power flows from the legitimacy of those institutions.'[136] In other words, the public is presented with a theater performance where elected officials pretend to be in control of national security decisions while the national security bureaucracies pretend to merely carry out the will of the elected government. This illusion tends to serve both sides well since elected officials cannot do much without the bureaucracies on their side and since open bureaucratic defiance of an elected government would smell like mutiny and thus erode the legitimacy of the whole government and, with it, also the source of the deep state's secret power.

There are some important features that make the existence of deep states relevant for 5GW. Deep states are both potential perpetrators of 5GW and potential targets of 5GW opponents. Deep states do not tend to be creative and they do not have any political objectives of their own, but, if any outside group was able to infiltrate them, they would be extremely well positioned for an attack against a state or a society. Throughout the Cold War, the Soviet Union made tremendous efforts to penetrate Western intelligence services, not just to obtain state secrets but also to shape the perceptions of Western governments.[137] Deep states can operate in secrecy, they often control tremendous financial and material resources, they sometimes have access to technological and weapons capabilities beyond the regular military and police forces, and they can reach deep into a society owing to surveillance and covert influence on the media, as well as other corporate and civil society actors.[138]

Any infiltration of a deep state can potentially enable control of a government's perceptions by way of fabricated or manipulated intelligence, which can influence policies. Foreign intelligence services and even terror groups have penetrated U.S. intelligence agencies, which has not only resulted in the leaking of highly sensitive national security information to enemies but has also enabled enemies to shape U.S. perceptions and to carry out attacks against American targets.[139] Attempts to recruit individuals from adversarial organizations open up an intelligence agency to being infiltrated by double agents.[140] During the Cold War, there were justified fears that Western intelligence could be penetrated at the highest level, with even the heads of MI5 and the CIA coming under suspicion.[141] Deep states are also vulnerable to co-optation by economic/corporate or criminal interests, as happened in Russia in the 1990s. Once an organization has been penetrated, a traitor could facilitate the recruitment and placement of more traitors, and over time traitors in key positions could take over the organization and abuse its powers to advance the objectives of a foreign adversary. The fundamental problem is that it is a function of the deep state (the intelligence services) to guard against infiltration and treason—but who guards the guardians and makes sure the guardians have not themselves succumbed to treason?

Conclusion

Fifth generation warfare is war at the societal level that relies heavily on psychological manipulation and minimizes the use of violence. This suggests not that 5GW would necessarily be nonviolent, but that violence is not the primary mechanism for coercing an enemy. The intellectual roots of 5GW are the concepts of netwar/noopolitik and also Qiao and Wang's unrestricted warfare theory, which emphasizes non-military over direct military means to gain an advantage. Where 5GW deviates from unrestricted warfare is in terms of its narrower focus on the manipulation of perception, culture, and cognition. Abbott moved the debate on 5GW forward by dropping the original theory of generational warfare in favor of gradients of warfare, which provides a better supporting logic for the emergence of 5GW. An important challenge to understanding 5GW is secrecy or, rather, how an adversary can manipulate the victim's perception of activities that are ultimately nefarious. In short, victims of 5GW usually do not believe they are under attack or do not understand from where an attack may be coming, which is the reason why they will be conquered. Finally, 5GW can be conducted by super-empowered individuals or small groups, who may work with criminal organizations and may be able to subvert states and, in particular, deep states to weaponize them against another group that they consider to be their enemy.

Notes

1 Thomas X. Hammes, *The Sling and the Stone: On War in the 21st Century* (St. Paul, MN: Zenith Press, 2004), 208.
2 Hammes, *The Sling and the Stone*, 290.
3 Paul Robinson, *Dictionary of International Security* (Cambridge: Polity, 2008), 223.
4 Carl von Clausewitz, *On War* (Princeton, NJ: Princeton University Press, 1984), 80–81.
5 Martin van Creveld, *The Transformation of War* (New York: Free Press, 1991), 126–149.
6 Thomas Rid, *Cyber War Will Not Take Place* (Oxford: Oxford University Press, 2013), 12.
7 The caveat here is that war would still need to coerce an opponent to achieve a desired outcome in a conflict. This coercion might not be physical but could still be considered violence in a broader meaning of violence, e.g. psychological violence or structural violence.
8 H.P. Albarelli, *A Terrible Mistake: The Murder of Frank Olson and the CIA's Secret Cold War Experiments* (Walterville, OR: TrineDay, 2009), 61–64.
9 David A. Koplov, *Non-Lethal Weapons: The Law and Policy of Revolutionary Technologies for the Military and Law Enforcement* (Cambridge: Cambridge University Press, 2006), 27–30.
10 U.S. Department of Defense, 'Joint Non-Lethal Weapons Program,' Joint Intermediate Force Capabilities Office, https://jnlwp.defense.gov/
11 U.S. Department of Defense, 'Department of Defense Directive Number 3000.03E,' April 25, 2013.
12 John B. Alexander, *Future War: Non-Lethal Weapons in the Twenty-First Century* (New York: St. Martin's Griffin, 1997), 16.
13 Robert Mandel, *Security, Strategy, and the Quest for Bloodless War* (Boulder, CO: Lynne Rienner, 2004), 16.
14 Samuel Moyn, *Humane: How the United States Abandoned Peace and Reinvented War* (New York: Farrar Strauss Giroux, 2021), 8.
15 Moyn, *Humane*, 6.
16 Moyn, *Humane*, 324.
17 John Arquilla and David Ronfeldt (eds.), *In Athena's Camp: Preparing for Conflict in the Information Age* (Santa Monica, CA: RAND, 1997).
18 Richard Szafranski, 'Neocortical Warfare? The Acme of Skill,' in: Arquilla and Ronfeldt, *In Athena's Camp*, 395.
19 Von Clausewitz, *On War*, 75.
20 Szafranski, 'Neocortical Warfare?' 398.
21 Szafranski, 'Neocortical Warfare?' 404.
22 John Arquilla and David Ronfeldt, 'The Advent of Netwar,' in: Arquilla and Ronfeldt (eds.), *In Athena's Camp*, 277.
23 Arquilla and Ronfeldt, 'The Advent of Netwar,' 281–282.
24 Arquilla and Ronfeldt, 'The Advent of Netwar,' 284.
25 John Arquilla and David Ronfeldt, *The Emergence of Noopolitik: Toward an American Information Strategy* (Santa Monica, CA: RAND, 1999).

26 Arquilla and Ronfeldt, *The Emergence of Noopolitik*, 13.
27 Arquilla and Ronfeldt, *The Emergence of Noopolitik*, 10.
28 Arquilla and Ronfeldt, *The Emergence of Noopolitik*, 29.
29 Arquilla and Ronfeldt, *The Emergence of Noopolitik*, 36.
30 Arquilla and Ronfeldt, *The Emergence of Noopolitik*, 45.
31 Arquilla and Ronfeldt, *The Emergence of Noopolitik*, 53.
32 Liang Qiao and Xiangsui Wang, *Unrestricted Warfare* (Naples, Italy: Albatross, 2020).
33 Robert Bunker, 'Unrestricted Warfare: Review Essay I,' *Small Wars and Insurgencies* 11, no. 1 (2000): 115.
34 Robert Spalding, *War without Rules: China's Playbook for Global Domination* (New York: Sentinel, 2022), XII.
35 Donald Reed, 'Beyond the War on Terror: Into the Fifth Generation of War and Conflict,' *Studies in Conflict & Terrorism* 31, no. 8 (2008): 690.
36 Michael Pillsbury, *The Hundred-Year Marathon: China's Secret Strategy to Replace America as the Global Superpower* (New York: St. Martin's Press, 2016), 214; Spalding, *War without Rules*, 8.
37 Qiao and Wang, *Unrestricted Warfare*, III–IV.
38 David Barno and Nora Bensahel, 'A New Generation of Unrestricted Warfare,' *War on the Rocks*, April 19, 2016, available at: https://warontherocks.com/2016/04/a-new-generation-of-unrestricted-warfare/
39 Spalding, *War without Rules*, 12.
40 Qiao and Wang, *Unrestricted Warfare*, III.
41 Qiao and Wang, *Unrestricted Warfare*, 32.
42 Qiao and Wang, *Unrestricted Warfare*, 5.
43 Qiao and Wang, *Unrestricted Warfare*, 18–20.
44 Qiao and Wang, *Unrestricted Warfare*, 38.
45 Qiao and Wang, *Unrestricted Warfare*, 39.
46 Joseph W. Sullivan, 'A BRICS Currency Could Shake the Dollar's Dominance,' *Foreign Policy*, April 24, 2023, available at: https://foreignpolicy.com/2023/04/24/brics-currency-end-dollar-dominance-united-states-russia-china/
47 Qiao and Wang, *Unrestricted Warfare*, 41.
48 Qiao and Wang, *Unrestricted Warfare*, 42.
49 Qiao and Wang, *Unrestricted Warfare*, 30.
50 Qiao and Wang, *Unrestricted Warfare*, 42.
51 Qiao and Wang, *Unrestricted Warfare*, 42.
52 Qiao and Wang, *Unrestricted Warfare*, 42.
53 Qiao and Wang, *Unrestricted Warfare*, 42.
54 Qiao and Wang, *Unrestricted Warfare*, 26.
55 Spalding, *War without Rules*, 103.
56 Qiao and Wang, *Unrestricted Warfare*, 42.
57 Qiao and Wang, *Unrestricted Warfare*, 42.
58 Robert Spalding, *Stealth War: How China Took Over while America's Elite Slept* (London: Portfolio/ Penguin, 2019), 114–115.
59 Qiao and Wang, *Unrestricted Warfare*, 42.
60 Qiao and Wang, *Unrestricted Warfare*, 42.
61 Spalding, *War without Rules*, 104.
62 Qiao and Wang, *Unrestricted Warfare*, 42.
63 Robert Bunker, *China's Securing, Shaping, and Exploitation of Strategic Spaces: Gray Zone Response and Counter-Shi Strategies* (Xlibris, Small Wars Pocket Book, 2019), 78–83.
64 Qiao and Wang, *Unrestricted Warfare*, 42–43.
65 Clive Hamilton and Mareike Ohlberg, *Hidden Hand: Exposing how the Chinese Communist Party Is Reshaping the World* (London: OneWorld, 2020), 228–231.

66 Qiao and Wang, *Unrestricted Warfare*, 43.
67 Douglas Guilfoyle, 'The Rule of Law and Maritime Security: Understanding Lawfare in the South China Sea,' *International Affairs* 95, no. 5 (2019): 999–1017.
68 Qiao and Wang, *Unrestricted Warfare*, 159.
69 Qiao and Wang, *Unrestricted Warfare*, 156.
70 Qiao and Wang, *Unrestricted Warfare*, 163.
71 Qiao and Wang, *Unrestricted Warfare*, 162.
72 Qiao and Wang, *Unrestricted Warfare*, 166.
73 Qiao and Wang, *Unrestricted Warfare*, 169.
74 'In strategy, the longest way round is often the shortest way home.' Basil Liddell-Hart, *Strategy* (New York: Penguin, 1991), 5.
75 The anthrax used in the Amerithrax attacks of 2001 was determined to have originated from a U.S. biodefense lab and likely involved a U.S. biodefense scientist with expert knowledge of anthrax production. The FBI claimed that the attacks were perpetrated by Bruce Ivins, who was a senior research scientist at Fort Detrick and specialized in anthrax vaccines and who had access to anthrax in his lab. See also: Graeme MacQueen, *The Anthrax Deception: The Case for a Domestic Conspiracy* (Atlanta, GA: Clarity Press, 2014).
76 Hammes, *The Sling and the Stone*, 290.
77 Peter Neumann, *Old and New Terrorism: Late Modernity, Globalisation and the Transformation of Political Violence* (Cambridge: Polity, 2009), 14–48.
78 Reed, 'Beyond the War on Terror,' 693.
79 Reed, 'Beyond the War on Terror,' 699–703.
80 Daniel H. Abbot (ed.), *The Handbook of 5GW* (Ann Arbor, MI: Nimble Books, 2010.
81 James Hasík, 'Beyond the Briefing: Theoretical and Practical Problems in the Works and Legacy of John Boyd,' *Contemporary Security Policy* 34, no. 3 (2013): 583–584.
82 William S. Lind, 'John Boyd's Art of War,' *The American Conservative*, August 16, 2013, available at: www.theamericanconservative.com/john-boyds-art-of-war/
83 Chad Kohalyk, '5GW as Netwar 2.0,' in: Abbott, *The Handbook of 5GW*, 40.
84 Daniel H. Abbott, 'Go Deep: OODA and the Rainbow of xGW,' in Abbott, *The Handbook of 5GW*, 175.
85 Abbott, 'Go Deep,' 178–179.
86 Kohalyk, '5GW as Netwar 2.0,' 42.
87 Daniel H. Abbott, 'The Terminology of XGW,' in Abbott, *The Handbook of 5GW*, 204–205.
88 Abbott, 'The Terminology of XGW,' 205.
89 Mark Safranski, 'Unto the Fifth Generation of War,' in Abbott, *The Handbook of 5GW*, 170.
90 Daniel McIntosh, 'Transhuman Politics and Fifth Generation War,' in Abbott, *The Handbook of 5GW*, 81.
91 Thomas X. Hammes, 'Fourth Generation Warfare Evolves, Fifth Emerges,' *Military Review* 87, no. 3 (2007): 20.
92 For a criticism of the inadequate understanding of history in 4GW, see Antulio Echevarria, 'Deconstructing the Theory of Fourth-Generation War,' *Contemporary Security Policy* 26, no. 2 (2005): 233–241.
93 Abbott, 'Introduction,' in Abbott, *The Handbook of 5GW*, 7.
94 Abbott, 'Introduction,' 8–9.
95 Abbott, 'Introduction,' 10.
96 Adam Herring, 'Searching for 5GW,' in: Abbott, *The Handbook of 5GW*, 71.
97 Herring, 'Searching for 5GW,' 75.
98 McIntosh, 'Transhuman Politics and Fifth Generation War,' 82.
99 L.C. Rees, 'The End of the Rainbow: Implications of 5GW for a General Theory of War,' in: Abbott, *The Handbook of 5GW*, 28.

100 Rory Cormac and Richard Aldrich, 'Grey Is the New Black: Covert Action and Implausible Deniability,' *International Affairs* 94, no. 3 (2018): 487–488.
101 Austin Carson, *Secret Wars: Covert Conflict in International Politics* (Princeton, NJ: Princeton University Press, 2018), 38.
102 Cormac and Aldrich, 'Grey Is the New Black,' 492.
103 John Earl Haynes and Harvey Klehr examined the argumentative strategies of academics on the left in relation to research indicating Soviet espionage in the United States. After the discovery of spies such as Alger Hiss, there was an effort first to deny their guilt and, when that was no longer possible, to downplay the consequences of their actions. It is a common cognitive error to cling to a world view even in the face of overwhelming evidence to the contrary. John Earl Haynes and Harvey Klehr, *In Denial: Historians, Communism & Espionage* (San Francisco, CA: Encounter Books, 2003).
104 Abbott, '5GW under the Microscope,' in Abbott, *The Handbook of 5GW*, 165.
105 Abbott, '5GW under the Microscope,' 165.
106 Shane Deichman, 'Battling for Perception: Into the 5th Generation?' in Abbott, *The Handbook of 5GW*, 15.
107 According to Paul Scharre, '[a] swarm consists of disparate elements that coordinate and adapt their movements in order to give rise to an emergent, coherent whole. A wolf pack is something quite different from a group of wolves. Ant colonies can build structures and wage wars, but a large number of uncoordinated ants can accomplish neither . . . These collective behaviors emerge because of simple rules at the individual level that lead to complex aggregate behavior . . . No individual ant "knows" which trail is fastest, but collectively the colony nonetheless converges on the optimal route.' Paul Scharre, 'Unleash the Swarm: The Future of Warfare,' *War on the Rocks*, March 4, 2015, available at: https://warontherocks.com/2015/03/unleash-the-swarm-the-future-of-warfare/
108 Irene Giardina, 'Collective Behavior in Animal Groups: Theoretical Models and Empirical Studies,' *HFSP J.* 2, no. 4 (2008): 205–219.
109 Stephen Pampinella, 'The Construction of 5GW,' in Abbott, *The Handbook of 5GW*, 55.
110 David Axe, 'Piracy, Human Security, and 5GW in Somalia,' in Abbott, *The Handbook of 5GW*, 152.
111 Thomas Rid, *Active Measures: The Secret History of Disinformation and Political Warfare* (New York: Farrar Strauss Giroux, 2020), 264–265.
112 Rid, *Active Measures*, 271–273.
113 Markus Wolf, *Memoirs of a Spymaster* (London: Pimlico, 1998), 242–243.
114 Samuel P. Huntington, *The Clash of Civilizations and the Re-Making of World Order* (New York: Simon & Schuster, 1996); Lind explored the cultural/civilizational dimension of future conflicts in his early work in relation to the cultural decline of the West and discussed the possibility of cultural wars. William S. Lind, 'Defending Western Culture,' *Foreign Policy* 84 (Autumn 1991): 40–50.
115 McIntosh, 'Transhuman Politics and Fifth Generation War,' 81.
116 Hammes, *The Sling and the Stone*, 291.
117 Thomas L. Friedman, *The Lexus and the Olive Tree: Understanding Globalization* (New York: Anchor Books, 2000), 433.
118 Although international relations literature has ignored, for a long time, the role of the super-wealthy in world politics, new research by Peter Hägel has shown that billionaires engage in transnational activism and that they do have enormous influence in several policy areas, including international security. Peter Hägel, *Billionaires in World Politics* (Oxford: Oxford University Press, 2020), 117–119.
119 Ian Sullivan, Matthew Santaspirt, and Luke Shabro, 'Mad Scientist Conference: Visualizing Multi Domain Battle 2030–2050,' Georgetown University, July 25–27, 2017, available at: https://community.apan.org/wg/tradoc-g2/mad-scientist/m/visualizing-multi-domain-battle-2030-2050/210183, 11.

120 McIntosh, 'Transhuman Politics and Fifth Generation War,' 76.
121 A famous example is the Medellín Cartel's war against the Colombian government from the mid-1980s to the early 1990s, during which the cartel killed thousands of police officers, journalists, judges, and government officials in order to change government policies on extradition.
122 U.S. Department of Justice, 'Overview of the Law Enforcement Strategy to Combat International Organized Crime,' April 2008, available at: www.justice.gov/archive/doj-espanol/speeches/2008/ioc-strategy-public-overview.pdf, 10.
123 Christopher White, *The War on Drugs in the Americas* (New York: Routledge, 2019), 99–102.
124 Robert Bunker and Pamela Bunker, 'Defining Criminal States,' *Global Crime* 7, no. 3–4 (2006): 375.
125 Yuriy A. Voronin, 'The Emerging Criminal State: Economic and Political Aspects of Organized Crime in Russia,' in Phil Williams, *Russian Organized Crime: The New Threat?* (London: Frank Cass, 1997), 53.
126 Voronin, 'The Emerging Criminal State,' 57.
127 Robert Mandel, *Dark Logic: Transnational Criminal Organizations and Global Security* (Stanford, CA: Stanford University Press, 2011), 24.
128 Some commentators have taken the more extreme view that the deep state is a conspiracy of senior government officials against the elected government to then dismiss this view as a political fantasy. David Rohde, *In Deep: The FBI, the CIA, and the Truth about America's 'Deep State'* (New York: W.W. Norton, 2020).
129 Paul Robinson, 'Military–Industrial Complex,' in: Paul Robinson, *Dictionary of International Relations* (Cambridge: Polity, 2008), 122.
130 Dwight D. Eisenhower, 'Farewell Address,' January 17, 1961.
131 Mike Lofgren, *The Deep State: The Fall of the Constitution and the Rise of a Shadow Government* (New York: Penguin Books, 2016), 32–34.
132 Sean McFate, *The New Rules of War: Victory in the Age of Durable Disorder* (New York: William Morrow, 2019), 158.
133 Ola Tunander, 'Democratic State vs. the Deep State: Approaching the Dual State of the West' in: Eric Wilson and Tim Lindsey (eds.), *Government of the Shadows: Parapolitics and Criminal Sovereignty* (London: Pluto Press, 2009), 56.
134 Michael J. Glennon, *National Security and the Double Government* (Oxford: Oxford University Press, 2015), 13.
135 Glennon, *National Security and the Double Government*, 26.
136 Glennon, *National Security and the Double Government*, 29.
137 Anatoly Golitsyn has detailed extensive KGB disinformation efforts that targeted both Western intelligence services and Western societies, one of which was the infamous Trust Operation of the 1920s. Anatoly Golitsyn, *New Lies for Old* (San Pedro, CA: GSG Associates, 1990), 10–17, 71.
138 A very good exploration of a deep state in an authoritarian system is Yevgenia Albats's book *State within a State: The KGB Past, Present and Future* (New York: Farrar Strauss Giroux, 1994).
139 Notable U.S. spy cases include former FBI agent Robert Hanssen, who was able to spy for the Soviet Union and Russia for over 20 years and who betrayed some of the most sensitive information related to U.S. nuclear strategy; former DIA analyst Ana Belén Montes, who spied for Cuba and manipulated U.S. perceptions of Cuba; and former Al-Qaeda operative Ali Mohamed, who penetrated the CIA, U.S. Special Forces, and the FBI and who provided Al-Qaeda with key information that enabled the 9/11 attacks.
140 What can be considered an early blunder of post-war U.S. intelligence was the recruitment of German military intelligence chief Reinhard Gehlen, who created the Gehlen Org on behalf of U.S. intelligence to provide intelligence on the Soviet Union in the aftermath of the war. The Gehlen Org was littered with Nazis and communist spies

who could influence U.S. perceptions of the Soviet Union and thereby U.S. foreign policy. Richard Breitman, Norman J. W. Goda, Timothy Naftali, and Robert Wolfe, *U.S. Intelligence and the Nazis* (Cambridge: Cambridge University Press, 2005).
141 The most famous case was Kim Philby, who was quite senior in MI6 before he was forced to leave the agency in 1955. David Wise, *Molehunt: The Secret Search for Traitors That Shattered the CIA* (New York: Random House, 1992), 102, 235.

References

Abbot, Daniel H. (ed.), *The Handbook of 5GW* (Ann Arbor, MI: Nimble Books, 2010).
———, 'Introduction,' in: Daniel H. Abbot (ed.), *The Handbook of 5GW* (Ann Arbor, MI: Nimble Books, 2010), 6–10.
———, '5GW under the Microscope,' in: Daniel H. Abbot (ed.), *The Handbook of 5GW* (Ann Arbor, MI: Nimble Books, 2010), 164–167.
———, 'Go Deep: OODA and the Rainbow of xGW,' in: Daniel H. Abbot (ed.), *The Handbook of 5GW* (Ann Arbor, MI: Nimble Books, 2010), 174–186.
———, 'The Terminology of XGW,' in: Daniel H. Abbot (ed.), *The Handbook of 5GW* (Ann Arbor, MI: Nimble Books, 2010), 204–205.
Albarelli, H.P., *A Terrible Mistake: The Murder of Frank Olson and the CIA's Secret Cold War Experiments* (Walterville, OR: TrineDay, 2009).
Albats, Yevgenia, *State within a State: The KGB Past, Present and Future* (New York: Farrar Strauss Giroux, 1994).
Alexander, John B., *Future War: Non-Lethal Weapons in the Twenty-First Century* (New York: St. Martin's Griffin, 1997).
Arquilla, John and Ronfeldt, David (eds.), *In Athena's Camp: Preparing for Conflict in the Information Age* (Santa Monica, CA: RAND, 1997).
———, 'The Advent of Netwar,' in: John Arquilla and David Ronfeldt (eds.), *In Athena's Camp: Preparing for Conflict in the Information Age* (Santa Monica, CA: RAND, 1997), 275–294.
———, *The Emergence of Noopolitik: Toward an American Information Strategy* (Santa Monica, CA: RAND, 1999).
Axe, David, 'Piracy, Human Security, and 5GW in Somalia,' in: Daniel H. Abbot (ed.), *The Handbook of 5GW* (Ann Arbor, MI: Nimble Books, 2010), 152–162.
Barno, David and Bensahel, Nora, 'A New Generation of Unrestricted Warfare,' *War on the Rocks*, April 19, 2016, available at: https://warontherocks.com/2016/04/a-new-generation-of-unrestricted-warfare/
Breitman, Richard, Goda, Norman J.W., Naftali, Timothy, and Wolfe, Robert, *U.S. Intelligence and the Nazis* (Cambridge: Cambridge University Press, 2005).
Bunker, Robert, 'Unrestricted Warfare: Review Essay I,' *Small Wars and Insurgencies* 11, no. 1 (2000): 114–121.
———, *China's Securing, Shaping, and Exploitation of Strategic Spaces: Gray Zone Response and Counter-Shi Strategies* (Xlibris, Small Wars Pocket Book, 2019).
Bunker, Robert and Bunker, Pamela, 'Defining Criminal States,' *Global Crime* 7, no. 3–4 (2006): 365–378.
Carson, Austin, *Secret Wars: Covert Conflict in International Politics* (Princeton, NJ: Princeton University Press, 2018).
Cormac, Rory and Aldrich, Richard, 'Grey Is the New Black: Covert Action and Implausible Deniability,' *International Affairs* 94, no. 3 (2018): 477–494.
Deichman, Shane, 'Battling for Perception: Into the 5th Generation?' in: Daniel H. Abbot (ed.), *The Handbook of 5GW* (Ann Arbor, MI: Nimble Books, 2010), 11–19.

Echevarria, Antulio, 'Deconstructing the Theory of Fourth-Generation War,' *Contemporary Security Policy* 26, no. 2 (2005): 233–241.

Eisenhower, Dwight D., 'Farewell Address,' January 17, 1961.

Friedman, Thomas L., *The Lexus and the Olive Tree: Understanding Globalization* (New York: Anchor Books, 2000).

Giardina, Irene, 'Collective Behavior in Animal Groups: Theoretical Models and Empirical Studies,' *HFSP J.* 2, no. 4 (2008): 205–219.

Glennon, Michael J., *National Security and the Double Government* (Oxford: Oxford University Press, 2015).

Golitsyn, Anatoly, *New Lies for Old* (San Pedro, CA: GSG Associates, 1990).

Guilfoyle, Douglas, 'The Rule of Law and Maritime Security: Understanding Lawfare in the South China Sea,' *International Affairs* 95, no. 5 (2019): 999–1017.

Hägel, Peter, *Billionaires in World Politics* (Oxford: Oxford University Press, 2020).

Hamilton, Clive and Ohlberg, Mareike, *Hidden Hand: Exposing how the Chinese Communist Party Is Reshaping the World* (London: Oneworld, 2020).

Hammes, Thomas X., *The Sling and the Stone: On War in the 21st Century* (St. Paul, MN: Zenith Press, 2004).

———, 'Fourth Generation Warfare Evolves, Fifth Emerges,' *Military Review* 87, no. 3 (2007): 14–23.

Hasík, James, 'Beyond the Briefing: Theoretical and Practical Problems in the Works and Legacy of John Boyd,' *Contemporary Security Policy* 34, no. 3 (2013): 583–599.

Haynes, John Earl and Klehr, Harvey, *In Denial: Historians, Communism & Espionage* (San Francisco, CA: Encounter Books, 2003).

Herring, Adam, 'Searching for 5GW,' in: Daniel H. Abbot (ed.), *The Handbook of 5GW* (Ann Arbor, MI: Nimble Books, 2010), 205–206.

Huntington, Samuel P., *The Clash of Civilizations and the Re-Making of World Order* (New York: Simon & Schuster, 1996).

Kohalyk, Chad, '5GW as Netwar 2.0,' in: Daniel H. Abbot (ed.), *The Handbook of 5GW* (Ann Arbor, MI: Nimble Books, 2010), 30–46.

Koplov, David A., *Non-Lethal Weapons: The Law and Policy of Revolutionary Technologies for the Military and Law Enforcement* (Cambridge: Cambridge University Press, 2006).

Liddell-Hart, Basil, *Strategy* (New York: Penguin, 1991).

Lind, William S., 'Defending Western Culture,' *Foreign Policy* 84 (Autumn 1991): 40–50.

———, 'John Boyd's Art of War,' *The American Conservative*, August 16, 2013, available at: www.theamericanconservative.com/john-boyds-art-of-war/

Lofgren, Mike, *The Deep State: The Fall of the Constitution and the Rise of a Shadow Government* (New York: Penguin Books, 2016).

MacQueen, Graeme, *The Anthrax Deception: The Case for a Domestic Conspiracy* (Atlanta, GA: Clarity Press, 2014).

Mandel, Robert, *Security, Strategy, and the Quest for Bloodless War* (Boulder, CO: Lynne Rienner, 2004).

———, *Dark Logic: Transnational Criminal Organizations and Global Security* (Stanford, CA: Stanford University Press, 2011).

McFate, Sean, *The New Rules of War: Victory in the Age of Durable Disorder* (New York: William Morrow, 2019).

McIntosh, Daniel, 'Transhuman Politics and Fifth Generation War,' in: Daniel H. Abbot (ed.), *The Handbook of 5GW* (Ann Arbor, MI: Nimble Books, 2010), 76–97.

Moyn, Samuel, *Humane: How the United States Abandoned Peace and Reinvented War* (New York: Farrar Strauss Giroux, 2021).

Neumann, Peter, *Old and New Terrorism: Late Modernity, Globalisation and the Transformation of Political Violence* (Cambridge: Polity, 2009).

Pampinella, Stephen, 'The Construction of 5GW,' in: Daniel H. Abbot (ed.), *The Handbook of 5GW* (Ann Arbor, MI: Nimble Books, 2010), 47–62.

Pillsbury, Michael, *The Hundred-Year Marathon: China's Secret Strategy to Replace America as the Global Superpower* (New York: St. Martin's Press, 2016).

Reed, Donald, 'Beyond the War on Terror: Into the Fifth Generation of War and Conflict,' *Studies in Conflict & Terrorism* 31, no. 8 (2008): 684–722.

Rees, L.C., 'The End of the Rainbow: Implications of 5GW for a General Theory of War,' in: Daniel H. Abbot (ed.), *The Handbook of 5GW* (Ann Arbor, MI: Nimble Books, 2010), 20–29.

Rid, Thomas, *Cyber War Will Not Take Place* (Oxford: Oxford University Press, 2013).

———, *Active Measures: The Secret History of Disinformation and Political Warfare* (New York: Farrar Strauss Giroux, 2020).

Robinson, Paul, *Dictionary of International Security* (Cambridge: Polity, 2008).

———, 'Military–Industrial Complex,' in: Paul Robinson, *Dictionary of International Relations* (Cambridge: Polity, 2008).

Rohde, David, *In Deep: The FBI, the CIA, and the Truth about America's 'Deep State'* (New York: W.W. Norton, 2020).

Safranski, Mark, 'Unto the Fifth Generation of War,' in: Daniel H. Abbot (ed.), *The Handbook of 5GW* (Ann Arbor, MI: Nimble Books, 2010), 169–174.

Scharre, Paul, 'Unleash the Swarm: The Future of Warfare,' *War on the Rocks*, March 4, 2015, available at: https://warontherocks.com/2015/03/unleash-the-swarm-the-future-of-warfare/

Spalding, Robert, *Stealth War: How China Took Over while America's Elite Slept* (London: Portfolio/Penguin, 2019).

———, *War without Rules: China's Playbook for Global Domination* (New York: Sentinel, 2022).

Sullivan, Ian, Santaspirt, Matthew, and Shabro, Luke, 'Mad Scientist Conference: Visualizing Multi Domain Battle 2030–2050,' Georgetown University, July 25–27, 2017, available at: https://community.apan.org/wg/tradoc-g2/mad-scientist/m/visualizing-multi-domain-battle-2030-2050/210183

Sullivan, Joseph W., 'A BRICS Currency Could Shake the Dollar's Dominance,' *Foreign Policy*, April 24, 2023, available at: https://foreignpolicy.com/2023/04/24/brics-currency-end-dollar-dominance-united-states-russia-china/

Szafranski, Richard, 'Neocortical Warfare? The Acme of Skill,' in: John Arquilla and David Ronfeldt (eds.), *In Athena's Camp: Preparing for Conflict in the Information Age* (Santa Monica, CA: RAND, 1997), 395–416.

Tunander, Ola, 'Democratic State vs. the Deep State: Approaching the Dual State of the West' in: Eric Wilson and Tim Lindsey (eds.), *Government of the Shadows: Parapolitics and Criminal Sovereignty* (London: Pluto Press, 2009), 56–72.

U.S. Department of Defense, 'Joint Non-Lethal Weapons Program,' Joint Intermediate Force Capabilities Office, https://jnlwp.defense.gov/, accessed August 9, 2023.

———, 'Department of Defense Directive Number 3000.03E,' April 25, 2013.

U.S. Department of Justice, 'Overview of the Law Enforcement Strategy to Combat International Organized Crime,' April 2008, available at: www.justice.gov/archive/doj-espanol/speeches/2008/ioc-strategy-public-overview.pdf

Van Creveld, Martin, *The Transformation of War* (New York: Free Press, 1991).

Von Clausewitz, Carl, *On War* (Princeton, NJ: Princeton University Press, 1984).

Voronin, Yuriy A., 'The Emerging Criminal State: Economic and Political Aspects of Organized Crime in Russia,' in Phil Williams, *Russian Organized Crime: The New Threat?* (London: Frank Cass, 1997).

White, Christopher, *The War on Drugs in the Americas* (New York: Routledge, 2019).

Wise, David, *Molehunt: The Secret Search for Traitors That Shattered the CIA* (New York: Random House, 1992).

Wolf, Markus, *Memoirs of a Spymaster* (London: Pimlico, 1998).

3 The Counterinsurgency Paradigm in 5GW

The main objective in 5GW is the exercise of influence or control over a population for the purpose of either societal destabilization, regime change, political subjugation, or all three. As previously established, a successive generation of warfare 'is a strategic innovation designed to defeat the previous one.'[1] The paradigm of 4GW is insurgency or political warfare that seeks to undermine the political will of an opposing government to continue the fight. Hence, if one accepts this logic, then the underlying strategic paradigm of 5GW must be counterinsurgency. Since 5GW operates within the human domain or a society, 5GW opponents must seek to achieve dominance in the human domain. The first section will review some of the key ideas of counterinsurgency doctrine, which will later be connected to 5GW as a domestic counterinsurgency in which a supra-combination of state and nonstate actors seeks to control the collective behaviors of a population, with the aim either of complete subjugation or of weaponizing individuals or groups against a state or another social group. The second part of the chapter will outline psychological warfare (psywar) and its primary techniques. The final part of the chapter deals with social engineering as a method for manipulating collective human behavior by way of social conditioning, nudges, and other behavioral interventions.

The Counterinsurgency Paradigm

A counterinsurgency campaign is characterized by a government seeking to suppress irregular forces such as guerrillas and terrorists.[2] In counterinsurgency theory, there are enemy-centric and population-centric strategies; the former aim to destroy the irregular forces while the latter focus on population control.[3] Typically, counterinsurgency campaigns must achieve both the military and the political defeat of insurgents. Arguably, the objective of undermining the insurgents politically, which is referred to as 'winning the hearts and minds of the population,' is more important than killing or capturing insurgents.[4] The reason is that, as long as the insurgents have a compelling cause that motivates their fight and a significant segment of the population believes in it, insurgent forces can be reconstituted by others.[5] Furthermore, the '[i]njudicious use of firepower creates blood feuds, homeless people, and societal disruption that fuels and perpetuates an insurgency.'[6] At the same time, counterinsurgency cannot be limited to political measures and

DOI: 10.4324/9781003396963-4

psychological operations for the simple reason that insurgents will engage in violent action, which necessitates a violent response. However, any use of violence by counterinsurgents must be highly targeted at enemy combatants and should not undermine the perception of safety of the local population.[7]

The French counterinsurgency theorist David Galula emphasized the political dimension in counterinsurgency, based on the French experience in the Algerian Civil War. Galula's work has remained influential to the present day, inspiring the 2006 *U.S. Army Counterinsurgency Manual* and the two U.S. counterinsurgency campaigns in Iraq and Afghanistan.[8] He argued that insurgents must be deprived of a compelling political cause that motivates their fight and that allows them to recruit others and secure their support for the struggle. Unfortunately, '[t]o deprive the insurgent of a good cause amounts to solving the country's basic problems.'[9] This may not always be possible as it may require a government to embrace political reforms that undermine its own power. As pointed out by Robert Taber, the British victory in its Malayan counterinsurgency campaign came at an 'extremely high [financial] cost' and it also 'hastened the independence of Malaya by some years.'[10] This means that, depending on the political circumstances—which include how much influence or control the counterinsurgent forces have over the government and also the attainability of some sort of political compromise that denies the insurgents a legitimate cause—counterinsurgency is more likely to fail than to succeed.

Waging War in the Human Domain

Counterinsurgency is, according to General Sir Rupert Smith, 'war amongst the people' and, hence, plays out in the human domain 'where our advantages of numbers and equipment are neutralized.'[11] This raises the question: What is a domain? The U.S. military distinguishes between five domains of warfare in which it conducts operations and produces effects, namely, land, sea, air, outer space, and cyberspace. According to Patrick Allen and Dennis Gilbert, a domain must have the following characteristics to function as a warfare domain:

1. 'Unique capabilities are required to operate in the domain.'
2. 'A domain is not fully encompassed by another domain.'
3. 'A shared presence of friendly and opposing capabilities is possible in the domain.'
4. 'Control can be exerted over the domain.'
5. 'A domain provides the opportunity for synergy with other domains.'
6. 'A domain provides the opportunity for asymmetric actions across domains.'[12]

Land and sea are the oldest domains. The air domain only emerged after the invention of powered flight, from 1904 onwards, and this led to the establishment of the U.S. Air Force in 1947.[13] The space domain necessitated the capability to leave the atmosphere and to operate in outer space, which became possible in the late 1950s and which eventually led to the weaponization of space from the 1980s and the formal establishment of a U.S. Space Force in 2019. Cyberspace was only

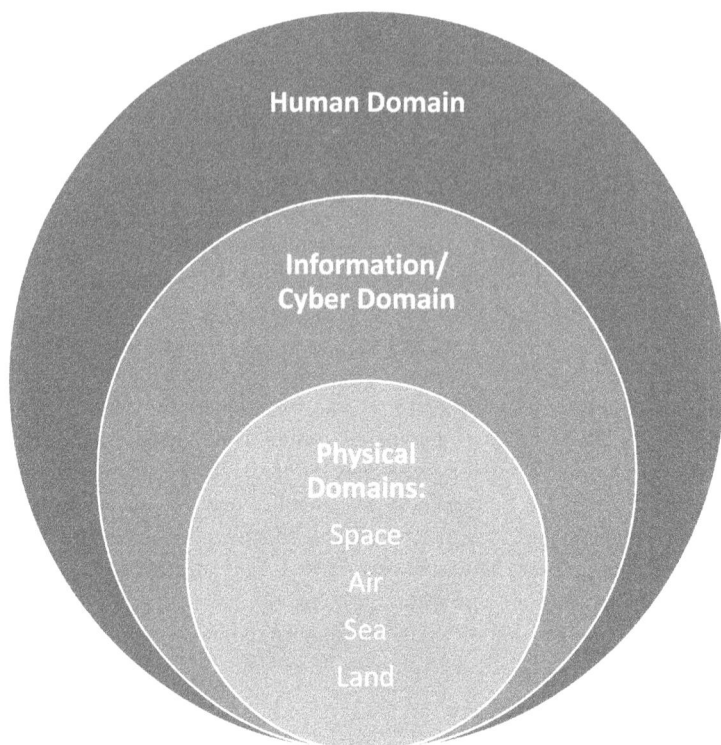

Figure 3.1 Domains of War
Source: Created by author

recognized as a separate domain in 2011 with the U.S. military establishing U.S. Cyber Command.[14] It should be noted that cyberspace is not a physical domain and it also overlaps with all of the older domains. Cyberspace is the realm of computers, computer networks, and information managed by the former.[15]

U.S. Special Operations Command proposed a sixth domain of warfare, which it called the 'human domain.' In a 2015 strategy paper, it defined the human domain as '[t]he people (individuals, groups, and populations) in the environment, including their perceptions, decision-making, and behavior.'[16] The paper further suggested that '[o]perations in the Human Domain depend on an understanding of, and competency in, the social, cultural, physical, informational, and psychological elements that affect and influence the domain.'[17] The people or a society are described as the medium in which insurgents and SOF operate. The objective is to shape the medium in a way that is conducive to achieving military and political goals.

Associated with the concept of the human domain is the notion of human terrain, which is defined as '[t]he social, ethnographic, cultural, economic, and political elements of the people with whom the Joint Force is operating.'[18] In other words, the human terrain is the operational environment for human domain operations, and it deals specifically with the structure of a society as defined in the ethnic composition, culture, and economic, political, and social relationships. The U.S. military tried to acquire better knowledge of the human terrain in Afghanistan and Iraq by developing the Human Terrain System, which put information collected by scientific Human Terrain Teams into a database that was available to military commanders. The idea was that, by developing cultural competence, U.S. counterinsurgent forces could operate more successfully in the human domain and could avoid the mistakes made in the Vietnam War.[19]

Separate Guerrillas from the Population

The human domain overlaps with the physical space or geography in which the population resides. Insurgents do not hold territory, but they need areas in which they can operate freely with the support of the population. Insurgents tend to operate in rural areas where there are fewer security forces and where there are fewer impediments to their movement. Operating in cities usually poses too great a risk for insurgents since they face far stronger opposition and it is easier for counterinsurgents controlling all the routes in and out of a city to trap them.[20] Therefore, Fidel Castro called cities 'the cemetery of revolutionaries and resources.'[21] Counterinsurgents are therefore usually in the unfortunate position of having to defend a population spread out over a large territory, which is being threatened by insurgent attacks and by political agitation and subversion.

Since insurgents hide within a population and receive critical support from a population, counterinsurgents must seek to physically separate the insurgents from the population.[22] Counterinsurgents must also seek to control the movement of the insurgents and prevent them from using neighboring states as safe havens or as supply routes. This means that the counterinsurgents must control roads (and other transportation routes) and the borders. Galula recommended that '[e]very inhabitant must be registered and given a foolproof identity card' for a census, which would be 'a source of intelligence as relationships, movement, and activities become easier to track.'[23] In a typical counterinsurgency campaign, a country is divided into areas controlled by counterinsurgents (white areas), areas controlled by insurgents (red areas), and areas that are currently contested (pink areas).[24] The basic strategy is usually summed up in the process of 'clearing, holding, and building': First, areas must be cleared of insurgents; second, the cleared areas need to be defended against any attempts by insurgents to return to these areas; and third, the counterinsurgents must work with the local population to build infrastructure and create permanent security.[25]

In countries that are large and have a lot of their population dispersed in rural areas with forests or jungle or mountains, insurgents have more opportunities to move freely and get local support, which puts the counterinsurgents at a massive disadvantage. In such a situation, it would be necessary to resettle a portion of the population and move it from the rural areas to more urban areas that are more easily defended and that enable greater control over a population that is not completely supportive of a government's cause.[26]

For example, part of the Briggs Plan in Malaya was the resettlement of 400,000 squatters into 400 new villages (1950–1951). These villages had barbed wire fences surrounding them, with floodlights and armed patrols, but they also contained schools, hospitals, and other facilities.[27] This approach to keeping guerrillas out of the villages was apparently so successful in Malaya that it was again attempted in Vietnam in 1962, under the name 'strategic hamlets.' However, owing to the immense corruption of the Diem regime, the strategic hamlet program was doomed from the start, as forced labor, confiscation, and repressive measures only alienated the rural population, which severely increased the hostility towards the regime.[28]

Undermine the Narratives and Political Objectives of the Adversary

Clausewitz recommended directing attacks against the enemy's center of gravity. In counterinsurgency, the center of gravity is the population, and the ultimate objective is the control of the population, as in the sense of shifting all support and loyalty from the insurgents to the government. In other words, the counterinsurgents must use a mix of propaganda and bribes (promises of political reform and economic benefits) to persuade the population to side with the government and not to support the insurgents. This means that the counterinsurgents must undermine the political cause of the insurgents, which gives them motivation and a measure of legitimacy. Without an attractive political cause, no insurgency is possible.[29] A good or effective cause is a liberation narrative, suitable for civil conflicts involving foreign occupation. Insurgents may also claim to fight the economic or political oppression of a particular societal group. Insurgents may use any kind of severe political problem to justify their violent struggle and may try to make things worse by violent action to demonstrate the inability of the government to deal with the problem effectively.[30]

The first task for the counterinsurgents is to gain the trust of the population by protecting the people against the insurgents. They must create a safe space for the indigenous population to earn their support. Once basic security is established, the counterinsurgents must counter the nationalist/liberation/ethnic/other narrative of the insurgents. It is widely understood that humans understand the world through narratives, which are stories that explain the world and give meaning to events.[31] They consist of protagonists (good guys), antagonists (bad guys), and a story arc (a logic with which events unfold). Insurgents and counterinsurgents are, hence, engaged in a battle of opposing narratives, which provide different interpretations of events, actions, and motivations. Counterinsurgents could either develop an alternative narrative to the insurgents' narrative or spin the insurgents' narrative in a way that it excludes the insurgents.[32] David Kilcullen suggested that one might 'use a nationalist narrative to marginalize foreign fighters in your area, or a narrative of national redemption to undermine former regime enemies that have been terrorizing the population.'[33]

Fifth Generation Warfare as a Counterinsurgency

The historical record of the failure of counterinsurgency campaigns does not suggest that 5GW was a losing proposition from the outset, since the actors and objectives in 5GW are likely to be very different from classical counterinsurgency. What is suggested here is not that 5GW would be equivalent to counterinsurgency, but merely that its strategic paradigm is derived from the historical practice of counterinsurgency. Just like in counterinsurgency, 5GW opponents fight in the human domain and depend on tactics and techniques of population control. Michael Flynn and Boone Cutler argued that 5GW is a 'war of narratives' and that '[f]ifth generation warfare evolved when the overlap of hybrid, irregular, and unrestricted warfare became directed at societies to affect the cognitive battlespace: the belief system of civilians.'[34]

Pampinella argued that 5GW as counterinsurgency 'succeeds by manipulating how insurgents and the indigenous populations perceive their own identity in relation to each other as well as to counterinsurgents.'[35] Furthermore, the security of the indigenous population would only be successfully achieved by empowering its members to protect themselves, which provides the counterinsurgents with an exit strategy.[36] Counterinsurgents practicing 5GW should undermine the insurgents' attempt to frame them as enemies by '1) taking on friendly role-identities and practicing self-restraint, and 2) altercasting potentially hostile Others with friendly role-identities and cooperative interests.'[37]

Brent Grace used the fight against a gang called Black Knights (BKs) that had dominated a housing complex in Chicago in the 1980s and early 1990s as an example of how counterinsurgency functions in 5GW.[38] The gang was successful in avoiding any crackdown by bribing local officials and keeping the violence at a low level. Since the police did not keep the local population safe, they were not trusted, and this meant that the gang could act as the authority within the area it controlled. As long as the government authorities did not care enough to intervene, the BKs could flourish. They were eventually defeated by a series of steps that increased the pressure on them and that radically altered the human terrain. This involved the creation of a new housing complex and the relocation of the population, with a new population of construction workers moving in.[39] The problem that the gang could not overcome was that the new population changed the culture and was far less vulnerable to the previously successful tactics of the BKs. This was accompanied by the prosecution of some of the gang leaders and a heavier police presence. In the end, the gang had to give up and leave the area as there was no path to victory left for it.[40]

Psychological Warfare

Fifth generation warfare is a war of deception and therefore a war of influence.[41] As in counterinsurgency, the use of psychological warfare is key to success. Unlike in classical counterinsurgency, psychological warfare is the primary weapon in 5GW as the use of force is de-emphasized. The target of attack, as well as the means to accomplish hidden objectives, is a population, which is to be manipulated into doing the bidding of a 5GW opponent. Psychological warfare is a broad term for any method that is used to psychologically manipulate an adversary. It is usually understood to be an auxiliary military activity in war that targets the enemy and seeks to reduce the effectiveness of enemy forces.[42] Part of psychological warfare is the use of propaganda or information that is distributed to enemy forces and populations by suitable media. However, propaganda is also used to influence the propagandist's own population (e.g. to sustain the political will to fight) and neutral societies (e.g. to encourage them to join a conflict on the side of the sponsor of propaganda or to abstain from a conflict), and it can be used both in wartime and peacetime to advance any political objective.

Functions and Approaches

Psychological warfare is the application of psychology to the problem of warfare in order to gain a military advantage.[43] According to a 1953 definition from the U.S. Joint Chiefs of Staff,

> [p]sychological warfare comprises the planned use of propaganda and related informational measures designed to influence the opinions, emotions, attitudes, and behavior of enemy or other foreign groups in such a way as to support the accomplishment of national policies and aims, or a military mission.[44]

Psychological warfare can be divided into strategic, tactical, and consolidation types.[45] Strategic psychological warfare tries to shape the international information space and, hence, goes beyond a war zone or target society. Tactical psychological warfare targets enemy forces on the battlefield and enemy civilian populations. Finally, consolidation psychological warfare operations seek to persuade local populations not to oppose occupation forces and to reassure them that they will receive aid and other assistance.

The purposes of psychological warfare are therefore to degrade the performance of enemy forces, lower the probability of resistance, and secure the cooperation of an occupied population. This can be accomplished in the following ways:

1. *Deception*: Trick the enemy into making mistakes such as wasting or misallocating resources. Deception can either try to suppress signals (concealment or camouflage) or create false signals (spoofing). It works best when the deception reinforces the enemy's existing preconceptions, perceptions, and beliefs. A classical military deception was the Trojan horse that tricked the Trojans into taking Greek warriors hidden inside a wooden horse into their city; the Greeks then opened the gates, facilitating the defeat of Troy.[46]
2. *Confusion*: Cause inaction or delay a response by sowing confusion about major events or developments in a situation where there is a lack of good information. One approach for creating confusion is to make authoritative false claims that contradict the claims of an enemy in a situation where evidence one way or the other is absent. Another approach is to create a diversion that shifts the attention of a target audience from the important to the insignificant. More ambitious is the attempt to confuse a target audience as to who the real enemy is by

redefining their group identity—for example, by invoking another identity than the national one, such as subgroup identity, religious identity, gender identity, or ideological identity.

3. *Discord*: Create or exploit divisions of any kind to diminish the enemy's ability to mobilize all its resources in a conflict or war. This includes creating distrust and tensions between allies, between a population and its leaders, or between different ethnic, racial, religious, or other groups within the enemy's society. For example, Soviet psychological warfare of the Cold War targeted certain disenfranchised minority groups in the United States that had legitimate grievances, such as black people and Jews, to make them believe that they were victims of racism or anti-Semitism in order to create tensions and incite violence.[47]

4. *Demoralization*: Convince the enemy that its situation is hopeless, that its leadership is corrupt, and that resistance is futile in order to instill defeatism and encourage surrender. This includes creating doubt about the justness of its cause, the motivations and competence of its leadership, its overall ability to win a conflict or war, and, ultimately, individuals' ability to survive. The objective of demoralization is to encourage surrender, desertion, or an unwillingness to take risks and fight hard.[48]

5. *Terror*: Induce extreme fear in order to cause paralysis or a panic reaction that makes the enemy more vulnerable to a subsequent attack or more likely to surrender or comply with demands. Sudden or drastic action and demonstration of cruel intent and power can induce terror. Deliberate atrocities and the spreading of rumors of atrocities can be used to terrorize an enemy and make surrender more likely. For example, the Mongols massacred populations of captured cities if they refused to surrender right away and also spread rumors of these atrocities to other cities they wanted to conquer.[49] The American 'shock and awe' doctrine is based on the idea of achieving a psychological effect by hitting the enemy hard, across the whole battlefield, in a temporally condensed air campaign to achieve quick surrender.[50]

6. *Conversion*: Convert neutrals or enemies to one's own cause, faith, or ideology, thus reducing the numbers of supporters the enemy has and increasing the numbers of one's own supporters.[51] Conversion has been a psywar approach in many religious wars, where captured individuals and populations were given a chance to convert to the faith of their captors or suffer all the consequences of being treated as an enemy. However, conversion remains relevant in modern times in relation to ideologies such as fascism and communism. For example, Nazi psywar managed to convert some segments of occupied populations to the Nazi cause and recruited them into its military, such as the SS Division Galicia, which was staffed by Ukrainian Nazi converts.[52]

7. *Consolidation*: Reassure enemy forces and populations that they will be treated well after surrender, and that it is in their best interest to cooperate with liberating forces. This includes messaging that emphasizes that surrender is not betrayal or treason, but is the reasonable and honorable thing to do, and that

individuals who have surrendered will be treated fairly and with dignity. Members of a liberated population need to be reassured that a better future awaits them, that the occupation is temporary, that they can regain independence, and that they have much to gain from cooperation with the occupiers.

The primary tools of psychological warfare are communication and the use of carefully designed messages that seek to persuade specific target audiences using a range of media such as radio, TV, movies, books, newspapers, leaflets, pamphlets, text messaging, as well as online media. In other words, psychological warfare is largely about propaganda.

Propaganda

The word propaganda is derived from the Latin *propagare*, which means to propagate or spread; hence, the purpose of propaganda is to spread ideas.[53] The word propaganda appeared in the English language in the early 19th century. Originally, it did not have the negative connotation that the term acquired later in the 20th century. Edward Bernays popularized the term in his 1928 book *Propaganda*. In his book, Bernays suggested that propaganda is a 'mechanism by which ideas are disseminated on a large scale . . . in the broad sense of an organized effort to spread a particular belief or doctrine.'[54] Propaganda has become a pejorative term owing to its association with Nazi propaganda, which means that, nowadays, governments and militaries prefer the use of more neutral terms, such as psychological operations (PSYOP), information operations (IO), strategic communication (SC), and military information support operations (MISO). When it comes to the basics, it is all still propaganda.

There are many definitions of propaganda, which tend to include that it is manipulative or persuasive speech, that it uses communications or mass media, and that it is intended to influence a target audience in a manner consistent with the goals of the sponsor of propaganda.[55] It is merely a tool for influencing larger audiences to affect their emotions, their thinking, and their behavior. The term propaganda is usually associated with the use of lies or untruths, but it is much more accurate to characterize propaganda as biased speech rather than untruthful speech. As suggested by Leonard Doob, propagandists are more concerned about persuasiveness than truthfulness, using truth or lies based on what they deem to be more persuasive or more effective for their purposes at a particular moment.[56] Propagandists prefer lying by spin and by omission since this is far less likely to tarnish the source of the propaganda than getting caught in outright lies. A propagandist must protect the credibility of a propaganda source, as this is a precondition for the long-term effectiveness of propaganda.

Propaganda can be overt or covert and it can be inserted into any kind of media content such as news, entertainment, or the arts. War propaganda is the easiest to identify as such. It seeks to persuade the population of the justness of the cause, to convince it that the sacrifices of war are necessary, to demonize or dehumanize the enemy (e.g. creating outrage over the alleged misdeeds or atrocities of the enemy), and to reassure it that its leadership is competent, its military is strong, and the war will be won.

The Role of Ideology

The relationship between propaganda and ideology is complicated. On the surface, propaganda does not need to contain ideology since it is more of a general technique of persuasion. However, it would be difficult for political propaganda to avoid ideology altogether, as political perception tends to be filtered through ideology. Naturally, U.S. State Department propaganda can be expected to present information from a liberal pro-democracy perspective, while Russian propaganda is likely to frame information in terms of its ideology of Neo-Eurasianism or neotraditionalism.[57]

A way to describe ideology is as a grand narrative or metanarrative that explains the political world and that guides or justifies political action. Maurice Cranston suggested that ideology, in its narrow political definition, has the following five characteristics.

(1) it contains an explanatory theory of a more or less comprehensive kind about human experience and the external world; (2) it sets out a program, in generalized and abstract terms, of social and political organization; (3) it conceives the realization of this program as entailing a struggle; (4) it seeks not merely to persuade but to recruit loyal adherents, demanding what is sometimes called commitment; (5) it addresses a wide public but may tend to confer some special role of leadership on intellectuals.[58]

Among the most influential political ideologies in modern history are liberalism, fascism, communism, Salafism, and globalism. Where ideology is relevant in psychological warfare is in the use of destructive ideologies to blow up a society from within. There are ideologies that are inherently destructive as they seek to incite revolution, incite violence against a particular group that is declared responsible for a political or religious wrong, or encourage behaviors that are self-destructive or destructive to a society as a whole in the long term. Destructive ideologies are not supposed to work as a genuine or legitimate political program. Their purpose is to create societal dysfunction and chaos, which is meant to facilitate polarization, radicalization, and political violence.

Marxism is undoubtedly a destructive ideology as it postulates that capitalism must be overthrown by way of violent revolution.[59] But, once a revolution does occur, the communist paradise can still not be achieved, as a dictatorship of the proletariat is needed to violently suppress reactionary groups.[60] The communist paradise then gets postponed indefinitely. The Soviet Union used ideological subversion offensively to destabilize and overthrow other societies in order to replace capitalist governments with communist dictatorships, relying on techniques such as rigged elections, political coups, and communist insurrection. Despite the collapse of the

Soviet Union, Marxism remains tremendously influential as an ideology since it promises the poor social justice or a redistribution of wealth and thereby plays one societal group (the poor or underclass) against another societal group (the middle class or bourgeoisie), as they are blamed for the misfortunes of the former.[61]

Psychological Warfare Techniques

Apart from the use of overt or white propaganda using mass media and overt psychological operations on the battlefield such as loudspeaker appeals and leaflet drops, there are also covert methods that seek to avoid attribution or that go beyond communication.

Gray Propaganda

This is propaganda that is typically produced and disseminated by front organizations, which publicly misrepresent their purpose and/or their funding sources. They may pose as independent or politically neutral organizations. Most of the communist propaganda that was directed against the West relied on front organizations such as the World Peace Council, the Afro-Asian People's Solidarity Organization, the Christian Peace Conference, and the World Federation of Democratic Youth.[62] All kinds of NGOs, charities, religious organizations, think tanks, and other political lobby groups have been created by governments to exercise covert influence and disseminate propaganda abroad.

Rumors

The British Political Warfare Executive (PWE) invented the use of rumors as a psywar technique during the Second World War. A rumor can be defined as 'information that lacks secure standards of evidence. In the case of claims spread by word-of-mouth or through underground media, they are accepted based on their plausibility, rather than through the announcement of an official source.'[63] Rumors can be spread by human agents or by the mass media. Rumors are particularly effective in situations where there is some kind of emergency and a lack of authoritative information.[64]

Social Media

The main mechanism for spreading rumors and influencing large numbers of people nowadays is social media, which rival the traditional media in terms of public attention.[65] Social media platforms allow individuals to create and publish content in the form of text, pictures, and videos that become available to millions of people around the world, with almost no cost associated with publication and distribution. Consumers of social media often consider this type of content to be more credible than official or mainstream media content because social media content appears to be grassroots, immediate, unfiltered, and authentic. This obviously makes it worthwhile for propagandists to create social media content by way of internet trolls (and bots for amplification). The trolls post propaganda messages, antagonize users to force those on the fence into a camp, create doubt and confusion, make any reasonable debate impossible, and deny others the ability to exercise influence on platforms. A report to the British Joint Threat Research and Intelligence Group suggested the following techniques in social media influence operations:

> uploading YouTube videos containing persuasive messages; establishing on-line aliases with Facebook and Twitter accounts, blogs and forum member-ships for conducting HUMINT or encouraging discussion on specific issues; sending spoof emails and text messages as well as providing spoof online resources; and setting up spoof trade sites.[66]

These techniques are used by both state and nonstate actors engaged in psychological warfare.

Disinformation

The term disinformation is often used carelessly and inaccurately in the sense of any kind of false information that distracts from the true or official information. However, there is a huge difference between 'misinformation,' which is false information coming from an official foreign source, and 'disinformation,' which is deliberate false information that is disseminated through trusted sources.[67] According to a CIA definition from 1965, '[disinformation] . . . is false, incomplete information, or misleading information that is passed, fed, or confirmed to a targeted group, or country.'[68] Typically, disinformation is backed up by fabricated evidence such as forgeries, false expert witness accounts, fabricated intelligence, or manipulated official investigations, and it is placed in media that a target audience trusts, e.g. by way of a bribed newspaper editor or journalist. An intelligence service may first place a story in a neutral or developing country in the hope that it will be picked up by the international press. Disinformation is sticky as it can remain influential long after false claims have been disproven.[69]

PSYACTs

According to a U.S. Army Psychological Operations manual, PSYACTs are

> [p]lanned psychological activities across the range of military operations directed at the civilian population located in areas under friendly control in order to achieve a desired behavior that supports the military objectives and the operational freedom of the supported commanders.[70]

Furthermore, it is stated that 'PSYACTs are operations, conducted by SOF and conventional forces or other agents of action, which are planned and conducted as part of a PSYOP supporting program' and include the following: raids, strikes, shows of force, demonstrations, insurgency operations, and civic action programs.[71] An example of a PSYACT is to distribute goods such as food to partners with a printed PSYOP message such as 'Gift from the United States' in order to influence partners' views of the United States.[72]

Black Propaganda

Black propaganda, which is propaganda disseminated by a falsely identified or fake source, provides a channel for disinformation designed to confuse and disorient an adversary rather than convince a target audience of some false narrative. Black propaganda is therefore more effective causing inaction in a target audience than causing a particular action such as surrender. It can be aimed at both an enemy population and the enemy's (internal) security apparatus. While black propaganda can include untruths, it is still much more challenging to produce because it cannot contain errors that undermine the falsely stated identity of the source; for example, if the source pretends to be an internal resistance group, it cannot contain linguistic or cultural mistakes or inaccurate information that the target audience can easily recognize as such.[73] Furthermore, black propaganda can be spread through sources that an audience is more likely to trust in a situation where there is an open military or political conflict with the sponsor of the propaganda, such as real or invented opposition groups. Black propaganda may also impersonate the enemy's official channels/media if this can be accomplished technically.

Counterpropaganda

Any sophisticated adversary can be expected to conduct some psychological warfare of its own, which means that, usually, one's own propaganda competes with enemy propaganda. The enemy's propaganda must be therefore countered in order to reduce potential negative effects on the morale of one's own forces and population. According to Linebarger, '[c]ounterpropaganda is designed to refute a specific point or theme of enemy propaganda.'[74] Similarly, Herbert Romerstein defined counterpropaganda as 'carefully prepared answers to false propaganda with the purpose of refuting the disinformation and undermining the propagandist.'[75] For example, if enemy propaganda makes an accusation of war crimes that were supposedly perpetrated by one's own side, then counterpropaganda must address the claim and must seek to undermine it with facts and other evidence that refute it. However, counterpropaganda is not simply about fact-checking; it is about persuasion and psychological manipulation. Like all propaganda, counterpropaganda will rely on true information as much as possible, will omit any information that contradicts favored narratives, and will use false information if such falsehoods are difficult to disprove.

For example, there has been a proliferation of fact-checking websites that claim to be independent arbiters of truth and advocates of reason in the face of online misinformation and disinformation. According to a dated Reuters survey, there were 113 fact-checking groups active globally in 2016, of which 40 percent were associated with news outlets, and 60 percent were independent groups or civil society organizations.[76] Many of the fact-checking websites receive foundation or corporate grants, which calls into question their proclaimed independence.[77]

The fact-checking sites either use a simplistic binary conception of truth where a claim is either true or false or they have a 'truth meter' or ranking system for claims. The general methodology is debunking, which means to refute claims by exposing bias, citing facts or expert claims that contradict the propaganda claims, or creating doubt about a claim's veracity (if it cannot be disproven outright), and ad hominem attacks against those who make certain claims. Usually, fact-checking websites will reference government or other official sources, which effectively amplifies government propaganda.

The fact-checking websites have become an important part of what Michael Schellenberger has called the 'censorship industrial complex.' In his testimony to the House Judiciary Select Subcommittee on the Weaponization of the Federal Government, Schellenberger argued that '[t]he intellectual leaders of the censorship complex have convinced journalists and social media executives that accurate information is disinformation, that valid hypotheses are conspiracy theories, and that greater self-censorship results in more accurate reporting.'[78] Obviously, by suppressing inconvenient information by inaccurately labeling it as misinformation or disinformation, misinformation and propaganda can prosper to the detriment of the truth and the interests of citizens.

Social Engineering

Social engineering is a term that describes the hacking of humans or individuals by exploiting knowledge from the behavioral and social sciences.[79] It overlaps substantially with psychological warfare and propaganda, but also includes a greater range of tools that can be employed for psychological manipulation at a societal level, such as incentives and coercion. Social engineering is mostly a state-like capability, but it can also be employed by corporations and civil society actors. According to Marie Jones and Larry Flaxman, social engineering is a

> tactic of politics, religion, and corporate consumerism, even education and academia, [and it] involves literally engineering the behavior, attitudes, and desires of large groups of people . . . [it] is a way to gradually and subtly manipulate, coerce, or influence a segment of the population. This is most effectively achieved on the subconscious level.[80]

Social engineering is not necessarily nefarious, as it could theoretically be employed to address societal problems such as crime, environmental degradation, obesity, addiction, or political radicalization.[81] However, the same social engineering techniques used to change collective attitudes and behaviors in a positive way can also be abused for the political manipulation of a social group or a population to embrace policies that are detrimental to individuals and even society as a whole. There are particular concerns about the use of social engineering for population control and political subjugation, which can amount to covert warfare waged by a corrupt power elite against a population. The main historical precedents can be found in the totalitarian regimes of the Soviet Union in the Stalin era, the Third Reich, and communist China under Mao.

Exploiting Human Nature

Human nature, reasoning, and decision-making are flawed, and these flaws are exploitable by social engineers. They include trust in authority, peer pressure, mass formation, confirmation bias, and cognitive dissonance, among others. Human behavior is somewhat predictable and, hence, can be influenced more easily once it is understood how an average person behaves in a particular or controlled situation. For example, in his famous experiments from 1961 and 1962, Stanley Milgram showed that most test subjects can be manipulated into torturing an innocent person.[82]

The experiment was set up as a scientist directing a teacher (the unwitting test subject) to facilitate learning by punishing a student (an actor) with (fake) electroshocks for wrong answers. Almost two-thirds (65 percent) of the test subjects were willing to administer increasingly higher electroshocks, even at the point where they had to assume that the student was unconscious and that higher electroshocks could be lethal. The test subjects willing to continue with the electroshocks to the maximum of 450 volts did generally voice concern to the scientist in charge, which proved that they understood the dangers, but this did not diminish their obedience to clearly unethical commands.[83] However, an important factor in average obedience was their perception of how legitimate the authority was.[84]

Another key experiment examining the role of peer pressure was developed earlier by psychologist Solomon Asch, in the 1950s. Asch set out to study the role of conformity and peer pressure in totalitarian systems of government by looking at how independent an individual's judgments were from a group.[85] The experiment worked as follows: A group of seven to nine participants was shown a figure with three vertical lines of different length and asked which of the lines was the longest. Unknown to the test subject, all of the other group members were instructed to give the same wrong answer and announce it one by one until the test subject was asked. The result was that only 24 percent of the test subjects consistently gave the obvious correct answers, while the answers of 27 percent of the test subjects were completely determined by the answers of the majority, and the remainder gave partially erroneous answers.[86] The experiment seems to indicate that only a quarter of test subjects are independent and confident enough to contradict a majority, while the majority will tend to at least publicly agree with the majority, even if the majority's position is obviously wrong, in order to avoid controversy.

The desire to conform with others is critical to mass formation, in which individuals are effectively acting under the hypnosis of a leader or authority, as argued by Mattias Desmet.[87] He postulated that there are four necessary conditions that lead to mass formation: (1) 'generalized loneliness, social isolation, and lack of social bonds among the population'; (2) 'lack of meaning in life'; (3) 'widespread presence of free-floating anxiety and psychological unease among a population'; and (4) 'a lot of free-floating frustration and aggression.'[88] In other words, mass formation requires a collective psychological crisis. Individuals become highly susceptible to narratives that are irrational but that create a new social bond, which is facilitated by ritualistic collective behaviors designed to create group cohesion.[89]

Once certain propaganda narratives are accepted, other cognitive biases make it extremely difficult for anybody subjected to mass formation to question even absurd narratives in the face of clear contradictory evidence. The most important of these biases is confirmation bias. In order to maintain their new beliefs, people will only seek out information that confirms their beliefs and will subconsciously ignore contradictory information. Leon Festinger argued that, once people have made a choice or formed an opinion, they are likely to stick with it, even when they encounter contradictory evidence. Contradictory information causes dissonance and discomfort, which people will naturally seek to reduce. Festinger reasoned, 'according to the theory the process of dissonance reduction should lead, after the decision, to an increase in the desirability of the chosen alternative and a decrease in the desirability of the rejected alternative.'[90] In other words, people who have bought into an obviously wrong propaganda narrative are likely to double down on their commitment to it because they cannot emotionally cope with dissonance.

Gaslighting is a social engineering technique that is designed to make people doubt their own perceptions of reality and make them accept reality as described by an authority. The term originates from the movie *Gaslight* from 1944, starring Charles Boyer and Ingrid Berman.[91] The story is about a young wife who is persuaded by her malicious husband that she is crazy. Strange things happen in the house, but the husband denies that they did, causing the confused wife to progressively doubt her own perceptions and her sanity. The mass media can gaslight a population by declaring obvious falsehoods as true, which will make a majority of people accept the falsehoods since they do not have enough confidence in their own perceptions and reasoning. However, for gaslighting to be successful, the victim's interpretation of reality must be so thoroughly undermined as to make a serious challenge to the abuser's false world view impossible.[92] People can be made to accept increasingly absurd interpretations of reality and, hence, be driven crazy. Whole societies can be driven crazy when official interpretations of reality and actual reality become irreconcilable. The bottom line is that the majority of people can be easily manipulated by authorities, propaganda, peer pressure, and lack of confidence in their own ability to make sense of reality.

Weaponized Education

The purpose of education is very simple to define—namely, prepare young people for life by equipping them with relevant skills and knowledge so that they can be successful and contribute to society. A secondary objective is to make them good citizens who follow the law and participate meaningfully and responsibly in politics. An education system becomes weaponized when it is used to indoctrinate and radicalize youth with a destructive ideology and when ideological indoctrination replaces education as its main purpose. Influence over a country's education system is key to any long-term plan for societal destabilization and totalitarian political control.

Young people are particularly vulnerable to ideological indoctrination as their brains are not fully developed and their minds lack the ability to think critically about information. Young brains are designed to absorb new information with ease, not to question information. Ideas that have been accepted during childhood will continue to influence the thinking of people into adulthood, as argued by Jacques Ellul. Ellul claimed that '[e]ducational methods play an immense role in political indoctrination . . . One must utilize the education of the young to condition them to what comes later.'[93] Research indicates that 'political values that are developed during schooling age tend to persist throughout life.'[94]

The dangers of societal radicalization through the education system are easily understood in the example of Mao's Cultural Revolution of the late 1960s. Mao manipulated the Chinese underclass to rebel against the bureaucratic clique in order to regain political control of China after the disastrous Great Leap Forward. Mao's strategy was to create disorder 'to achieve great order.'[95] He radicalized the youth by taking over the schools and universities using the Red Guards, who conducted a revolt against the education system.[96] The original Red Guards were leftist students from Tsinghua University Secondary School who denounced and terrorized suspected counterrevolutionaries and 'rightists.' They were instrumental in purging rightist (pro-capitalist) teachers, professors, and administrators, which allowed the 'anti-rightist campaigns to spread through China like wildfire.'[97]

A further consequence of an education system's shift from teaching important skills and useful knowledge to a focus on indoctrination is poor educational outcomes in terms of a failure of students to acquire even basic literacy and numeracy skills after eight or ten years of schooling.[98] In this context, the recently measured decline in IQ among people 18–22 years old by up to 3.8 IQ points, in several Western countries, is alarming, especially as IQ is strongly correlated to the quality of education.[99] Societal groups that favor more political control will, hence, undermine an education system by reducing educational standards and outcomes.[100] It is obvious that a dumbed-down population is more susceptible to government propaganda, is less inclined to insist on its rights, and is, hence, more controllable.[101]

Social Conditioning

A fundamental approach in social engineering is social conditioning, which is based on the research of Ivan Pavlov who demonstrated that one stimulus can be replaced by a random other stimulus to trigger a reflex behavior. According to Edward Hunter, Lenin summoned Pavlov to the Kremlin to explain his experiments on conditioning dogs in relation to how conditioning could be used on humans to create the 'new Soviet man.'[102] Conditioning is based on a simple stimulus–response model where a specific stimulus will trigger a specific behavioral response, as described in the behaviorism theories of John B. Watson and B.F. Skinner. Behaviorism postulates that all behavior results from external stimuli, making the study of the mind irrelevant. This idea is consistent with Marxism insofar as Marxism also denies the very existence of free will since everything must be a function of material conditions.[103] Conditioning was used alongside indoctrination as a key aspect of communist brainwashing to make people obedient to the Soviet regime.[104]

In the 1960s, the basic theory of conditioning was applied by the Soviets to military operations and foreign policy under the descriptive term of 'reflexive control,' based on information warfare means.[105] A NATO Research Paper by Can Kasapoglu defines reflexive control as 'the systematic methods of shaping the adversary's perceptions, thereby decisions, and latently forcing him to act voluntarily in a way that would be favourable to Russia's interests.'[106] Timothy Thomas suggests that

> reflexive control occurs when the controlling organ conveys (to the objective system) motives and reasons that cause it to reach the desired decision, the nature of which is maintained in strict secrecy. The decision itself must be made independently. A 'reflex' itself involves the specific process of imitating the enemy's reasoning or imitating the enemy's possible behavior and causes him to make a decision unfavorable to himself.[107]

Reflexive control targets the enemy's decision-making with information that is designed to trigger a specific response that can be exploited. A reflex is an automatic response and, hence, not a conscious or rational decision.

Thomas includes the following reflexive control techniques: '[C]amouflage (at all levels), disinformation, encouragement, blackmail by force, and the compromising of various officials and officers.'[108] Deception, disinformation, and own actions are used in view of shaping the enemy's response and thereby undermining the enemy's ability to make good decisions for itself. In other words, reflexive response requires thinking several steps ahead to move the enemy's decision-making in a predetermined direction. The approach is primarily a manipulation of an adversary's perception and decision-making algorithm. Kasapoglu claimed that Russia's conquest of Crimea was aided by reflexive response—at the strategic, operational, and tactical levels—with Russia confusing the West about the nature, extent, and limits of its military actions to cause inaction or paralysis.[109] Obviously, the basic principle of reflexive control can be applied to a state's own population to reinforce propaganda and also for behavioral control.

Nudging

Nudging is a social engineering approach that was developed by Richard Thaler and Cass Sunstein. Thaler and Sunstein pointed out that their research resonated with many governments that have since created nudge units, including the United States, the United Kingdom, 'Australia, New Zealand, Germany, Canada, Finland, Singapore, the Netherlands, France, Japan, India, Qatar, and Saudi Arabia.'[110] A key assumption of nudging is that human behavior is predictable and can be influenced in a number of ways, such as how choices are presented, including the order in which they are presented. One example is anchoring: People will use the first option as the basis for making a choice. When people are given three choices for giving a tip, presented as 15 percent, 20 percent, and 25 percent, they will naturally gravitate towards the middle, or 20 percent, but, if no tip suggestion is given, they would typically only give 15 percent.[111]

Thaler and Sunstein argue that there is no way of avoiding exercising influence on people's choices since the way people choose is hardwired into them.[112] So, if one presented choices randomly, one would still end up steering people towards a *particular* random choice. Therefore, it would make more sense to at least steer people towards choices that are objectively best for them based on a scientific understanding of the choices, according to Thaler and Sunstein. For example, many people instinctively choose foods that are unhealthy and end up obese and with health conditions that result from obesity. Hence, governments could nudge them to make healthier choices for themselves.

Nudging is choice architecture, or a method of designing the presentation of choices. Thaler and Sunstein claim that 'the choice architecture . . . alters people's behavior in a predictable way without forbidding any option or significantly changing their economic incentives.'[113] If one wants to steer people towards a particular choice, one can make it a default option and create hurdles for people to opt out of the default option. Out of sheer convenience, most people will end up choosing the default option, which is supposedly the best option for them. Choices can be also framed in a way to make the preferred option look more attractive by emphasizing a positive outcome instead of a negative one—for example, a medical doctor might present the risk of a treatment in terms of the percentage of survivors rather than the percentage of deaths.[114]

While Thaler and Sunstein's theory sounds benign enough, there are also some obvious problems with the idea that a choice architect always knows what the best choice for an individual is and that they would put always the interests of the individual ahead of the interests of the choice architect. An obvious example of how nudges can be used to the detriment of the chooser concerns the privacy and community policies of social media platforms. Potential users are given the default choice of agreeing to the policies, which are spelled out in long sections of fine print, or disagreeing with them, which means not being able to use the platform at all. Nobody ever reads the policies as they are written in legalese and are, hence, not comprehensible to ordinary users. This allows the social platforms to slip in conditions that users might think twice about before agreeing to them if they were aware of them.[115] As a result, people now get constantly nudged to make choices that are decidedly not in their best interests.

The Behavior Change Wheel

Another behavioral science approach, which was used to ensure compliance with public health policies in the pandemic era, was the 'behavior change wheel' developed by Susan Michie, Maartje van Stralen, and Robert West, first proposed in 2011.[116] It starts with defining the desirable behavior change in order to design an intervention that is suitable for bringing about the change. Important prerequisites for the intervention are the existence of the skills needed to perform the desired be-havior, the existence of environmental constraints, and the existence of a sufficient intention. These can be summed up as capability, opportunity, and motive: It must be possible for an individual to perform the desired behavior, there has to be an opportunity to do so, and there has to be an incentive to do so.[117]

The potential interventions include education, persuasion, incentivization, co-ercion, training, restriction, environmental restructuring, modelling, and enabling. A researcher can use surveys to collect information from a relevant group or popu-lation to determine what it would take for a behavior to change, which helps in the intervention design. At the beginning of an intervention, the motivation is typi-cally lowest, which means that the focus must be on capability and opportunity. Over time, the desired behavior becomes a habit, and capability and opportunity become less important.[118] In other words, over time, people can be conditioned to automatically perform a new, desirable behavior without having to be reminded or incentivized to do so.

Co-optation of Science and the Media

Fifth generation warfare usually relies on the manipulation of proxies or the co-optation of autonomous actors. A government or another powerful actor co-opts scientists, corporations, and civil society organizations by creating incentives (pos-itive or negative) for them to promote desired behavioral changes. This is usually done through grants, tax benefits, regulatory measures, selective bailouts, intimi-dation, and lawfare. It is a sophisticated form of corruption that remains hidden from any casual observer and can affect societal watchdog organizations such as regulatory agencies.[119]

Government social engineering programs are generally declared to be based on science, which means that scientific studies have to be produced as a foundation for the promotion of collective behavioral change. It would be naïve to assume that scientists would be completely unbiased or that their findings would necessarily be true in an objective sense. The history of science has shown that many scien-tific theories eventually get overturned by new empirical findings.[120] It has even been suggested that most scientific research findings are false owing to bias and poor methodology.[121] According to a survey published in *Nature*, over 70 percent of researchers could not reproduce the findings of published research, and only 47 percent of reproduced research matched the original findings, which calls into question the validity of the findings that could not be reproduced.[122] Another survey studying scientific fraud found that 2 percent of respondents admitted to having falsified results, 14 percent had observed others falsifying results, and 72 percent had observed questionable practices and methods.[123]

Furthermore, scientists' careers depend on publications and grants, and most of the grants come from the government. Hence, scientists will research topics that are likely to be funded and will adjust their findings to be consistent with the objectives or interests of a sponsor to ensure future grants.[124] The great majority of scientist are politically left leaning, which makes them biased with respect to certain policies.[125] Most importantly, science is merely the application of the scientific method to a problem and it offers no guarantee of finding an ultimate truth. Science should not be dependent on a scientific consensus based on a given scientific paradigm, as truth does not depend on how many people happen to believe it, even if they are experts.[126]

What is important in social engineering, apart from using 'science' as a justification for social programs, is control of the mass media, which can disseminate propaganda and suppress information that contradicts preferred narratives. Again, the assumption of media independence is as wrong as the assumption of an independent and politically unbiased science community. More obvious is corporate influence on the mass media since media corporations depend largely on advertising money to stay in business and, hence, will not publish information that undermines the interests of their biggest advertising clients.[127]

A concerning trend is the increasing hidden government censorship of the media across Western nations. Although a democratic government has limited ability to financially incentivize media corporations, it is quite apparent that there is some kind of collusion between mass media and the government, as was revealed in relation to certain propaganda narratives and social programs of recent years.[128] The U.S. government has also flagged social media comments and accounts deemed to be disinformation in order to pressure social media platforms to take them down.[129] There was a controversial initiative by the Biden administration in May 2022 to establish a Disinformation Governance Board under the Department of Homeland Security to '"prevent" the circulation of disinformation,' which has raised concerns about government censorship of the media and social media.[130] During the pandemic era, the mass media were promoting narratives that were remarkably consistent with government policies concerning lockdowns, mask mandates, and of course mass vaccination, which looks like close coordination between the government and the media, not just nationally but internationally or worldwide.

Conclusion

The paradigm of 5GW is counterinsurgency, which seeks to manipulate the identity of a political or social group to change the context in which resistance against a government or other authority takes place. In classical counterinsurgency, the insurgents must be separated from the population, their movements must be restricted, and their narratives must be undermined. Psychological warfare and propaganda are the primary tools for affecting the perceptions, thinking, and behaviors of a population. The general idea is that, *on average*, human behavior is predictable, and that a better understanding of human behavior acquired through the behavioral sciences (psychology, sociology, and anthropology) makes it possible to steer the collective behavior of a population in one direction or another. This can be done for the purposes of demoralization (destroying the will to resist aggression), political destabilization (inducing societal chaos), or consolidating political control over a society (political subjugation). Fifth generation warfare is much more complex than a classical counterinsurgency approach, as the reality of the existence of a conflict or war may not even be apparent to the majority of a population. Furthermore, a government (or particular government agencies or particular civil society organizations) could be co-opted by 5GW belligerents, making it more difficult for a population to understand that they are under attack or from where the attack comes. For most people, the idea that attacks could originate from their own government, or corrupted elements thereof, is completely alien to them, which makes them easy victims in political systems that are sliding into totalitarianism, and it also makes them potential collaborators in their own subjugation or demise.

Notes

1 Stephen Pampinella, 'The Construction of 5GW,' in: Abbott (ed.), *The Handbook of 5GW*, 50.
2 David H. Ucko, 'Whither Counterinsurgency: The Rise and Fall of a Divisive Concept,' in: Paul B. Rich and Isabelle Duyvesteyn (eds.), *The Routledge Handbook of Insurgency and Counterinsurgency* (London: Routledge, 2012), 68.
3 Ucko, 'Whither Counterinsurgency,' 69.
4 Research by Christopher Paul et al. indicated that most COIN campaigns that were more population-centric succeeded (11 out of 15) while most enemy-centric campaigns failed (17 out of 44 successes). Christopher Paul, Colin P. Clarke, Beth Grill, and Molly Dunigan, 'Moving Beyond Population-centric vs. Enemy-centric Counterinsurgency,' *Small Wars & Insurgencies* 27, no. 6 (2016): 1028.
5 David Galula, *Counterinsurgency Warfare: Theory and Practice* (New York: Frederick E. Praeger, 1964), 77
6 David Kilcullen, *Counterinsurgency* (Oxford: Oxford University Press, 2010), 30.
7 Kilcullen, *Counterinsurgency*, 4.
8 Paul, Clarke, Grill, and Dunigan, 'Moving Beyond Population-centric vs. Enemy-centric Counterinsurgency,' 1021.
9 Galula, *Counterinsurgency Warfare*, 67.
10 Robert Taber, *The War of the Flea: The Classic Study of Guerrilla Warfare* (Lincoln, NE: Potomac Books, 2002), 140, 142.
11 Rupert Smith, *The Utility of Force: The Art of War in the Modern World* (London: Penguin, 2005), 276.
12 Patrick D. Allen and Dennis P. Gilbert, 'The Information Sphere Domain Increasing Understanding and Cooperation,' NATO Cooperative Cyber Defence Centre of Excellence (2018), 3.
13 Allen and Gilbert, 'The Information Sphere Domain Increasing Understanding and Cooperation,' 1.
14 U.S. Department of Defense, 'Department of Defense Strategy for Operating in Cyberspace,' July 2011, available at: https://csrc.nist.gov/CSRC/media/Projects/ISPAB/documents/DOD-Strategy-for-Operating-in-Cyberspace.pdf
15 Derek Reveron (ed.), *Cyberspace and National Security: Threats, Opportunities, and Power in a Virtual World* (Washington, DC: Georgetown University Press, 2012), 5.
16 U.S. Special Operations Command, 'Operating in the Human Domain,' Version 1.0, August 3, 2015, 76.
17 U.S. Special Operations Command, 'Operating in the Human Domain,' 76.
18 U.S. Special Operations Command, 'Operating in the Human Domain,' 77.
19 Ben Connable, 'Human Terrain System Is Dead . . . What? Building and Sustaining Military Cultural Competence in the Aftermath of the Human Terrain System,' *Military Review* (January–February 2018): 30.
20 The famous example is the Battle of Algiers (1956–1957), where paratroopers closed all exits from a city quarter, which trapped the guerrillas and resulted in 5,000 arrests.

21 Anthony King, 'Urban Insurgency in the Twenty-First Century: Smaller Militaries and Increased Conflict in Cities,' *International Affairs* 98, no. 2 (2022): 616.
22 Galula, *Counterinsurgency Warfare*, 77.
23 Galula, *Counterinsurgency Warfare*, 116–117.
24 Galula, *Counterinsurgency Warfare*, 81.
25 David H. Ucko, 'The Five Fallacies of Clear–Hold–Build: Counterinsurgency, Governance and Development at the Local Level,' *The RUSI Journal* 158, no. 3 (2013): 54–61.
26 Galula, *Counterinsurgency Warfare*, 111–112.
27 John Nagl, *Learning to Eat Soup with a Knife: Counterinsurgency Lessons from Malaya and Vietnam* (Chicago, IL: University of Chicago Press, 2005), 75.
28 Gerard DeGroot, *A Noble Cause? America and the Vietnam War* (London: Longman, 2000), 75–76.
29 Galula, *Counterinsurgency Warfare*, 18.
30 Taber, *The War of the Flea*, 19.
31 Cody D. Delistraty, 'The Psychological Comforts of Storytelling,' *The Atlantic*, November 2, 2014, available at: www.theatlantic.com/health/archive/2014/11/the-psychological-comforts-of-storytelling/381964/
32 Kilcullen, *Counterinsurgency*, 42.
33 Kilcullen, *Counterinsurgency*, 42.
34 Michael Flynn and Boone Cutler, *The Citizen's Guide to Fifth Generation Warfare* (Resilient Patriot, 2022), 1.5.
35 Pampinella, 'The Construction of 5GW,' 51.
36 Pampinella, 'The Construction of 5GW,' 51.
37 Pampinella, 'The Construction of 5GW,' 55.
38 Brent Grace, 'The War for Robert Taylor,' in: Abbott (ed.), *The Handbook of 5GW*, 136–151.
39 Grace, 'The War for Robert Taylor,' 149.
40 Grace, 'The War for Robert Taylor,' 149–150.
41 L.C. Rees, 'The End of the Rainbow: Implications of 5GW for a General Theory of War,' in: Abbot (ed.), *The Handbook of 5GW*, 24.
42 Paul M.A. Linebarger, *Psychological Warfare* (Eastford, CT: Martino Fine Books, 2022), 37.
43 Linebarger, *Psychological Warfare*, 24.
44 Quoted from Linebarger, *Psychological Warfare*, 277.
45 Stephen E. Pease, *Psywar: Psychological Warfare in Korea, 1950–1953* (Harrisburg, PA: Stackpole Books, 1992), 7–9.
46 Wilbur Schramm, Daniel Katz, Willmoore Kendall, and Thedore Vallance, 'The Nature of Psychological Warfare,' Research Analysis Corporation, McLean, VA, Technical Memo, April 23, 1954, 3.
47 Thomas Rid, *Active Measures: The Secret History of Disinformation and Political Warfare* (New York: Farrar Strauss Giroux, 2020), 134–141; U.S. Department of State, 'Active Measures: A Report on the Substance and Process of Anti-U.S. Disinformation and Propaganda Campaigns,' August 1986, 54.
48 Linebarger, *Psychological Warfare*, 211.
49 David Nicolette, *The Mongol Warlords: Genghis Khan, Kublai Khan, Hulegu, Tamerlane* (Poole, UK: Firebird, 1990), 21; Linebarger, *Psychological Warfare*, 14–16.
50 Harlan K. Ullman and James P. Wade, *Shock and Awe: Achieving Rapid Dominance* (Washington, DC: National Defense University, 1996).
51 Linebarger, *Psychological Warfare*, 46.
52 The Nazis recruited many volunteers to form non-German ethnic military formations from occupied territories, but the SS Division Galicia is one of the most famous. Basyl Dmytryshyn, 'The SS Division "Galicia": Its Genesis, Training, Deployment,' *Nationalities Papers* 21, no. 2 (1993): 53–73.
53 Garth S. Jowett and Victoria O'Donnell, *Propaganda and Persuasion* (Thousand Oaks, CA: Sage, 2012), 2.

54 Edward Bernays, *Propaganda* (New York: Horace Liveright, 1928), 20.
55 Terence H. Qualter, *Propaganda and Psychological Warfare* (Burtyrki Books, 2020), 44–47.
56 Leonard W. Doob, 'Goebbels' Principles of Propaganda,' *The Public Opinion Quarterly* 14, no. 3 (1950): 428.
57 Russia's primary philosopher of Neo-Eurasianism is Alexander Dugin, whose ideas regarding traditionalism and the civilizational conflict between Antlanticists and Eurasianists have been adopted by the Kremlin as some sort of unofficial Russian ideology set in opposition to Western liberalism. See Michael Millerman, *Inside 'Putin's Brain': The Political Philosophy of Alexander Dugin* (Millerman School, 2022).
58 Maurice Cranston, 'Ideology,' *Encyclopedia Britannica*, available at: www.britannica.com/topic/ideology-society, accessed on May 30, 2023.
59 Friedrich Engels and Samuel Moore (eds.), *Manifesto of the Communist Party. By Karl Marx and Friedrich Engels. Authorized English Translation* (Chicago, IL: Charles H. Kerr, 1888).
60 W. Cleon Skousen, *The Naked Communist* (Salt Lake City, UT: Izzard Ink, 2014), 53–55.
61 Stuart Jeffries, 'Why Marxism Is on the Rise Again,' *The Guardian*, July 4, 2012, available at: www.theguardian.com/world/2012/jul/04/the-return-of-marxism
62 Richard Shultz and Roy Godson, *Dezinformatsia: Active Measures in Soviet Strategy* (Washington, DC: Pergamon Brassey's, 1984), 118.
63 Marc Argemi and Gary Alan Fine, 'Faked News: The Politics of Rumour in British World War II Propaganda,' *Journal of War & Culture Studies* 12, no. 2 (2019): 177.
64 Qualter, *Propaganda and Psychological Warfare*, 117–118.
65 Mason Walker and Katerina Eva Matsa, 'News Consumption across Social Media in 2021,' Pew Research Center, September 20, 2021, available at: www.pewresearch.org/journalism/2021/09/20/news-consumption-across-social-media-in-2021/
66 Manddep K. Dhami, 'Behavioural Science Support for JTRIG's (Joint Threat Research and Intelligence Group's) Effects and Online HUMINT Operations,' Human Systems Group, Information Management Department, DSTL, March 10, 2011, 2.
67 Ion Mihai Pacepa and Ronald Rychlak, *Disinformation: Former Spy Chief Reveals Secret Strategies for Undermining Freedom, Attacking Religion, and Promoting Terrorism* (Washington, DC: WND Books, 2013), 35–36.
68 Shultz and Godson, *Dezinformatsia*, 37.
69 Kent Clizbe, *Willing Accomplices: How KGB Covert Influence Agents Created Political Correctness, Obama's Hate America First Political Platform, and Destroyed America* (Ashburn, VA: Andemca, 2011), 97.
70 U.S. Department of Defense, 'FM3–05.301: Psychological Operations Tactics, Techniques, and Procedures,' Department of the Army, December 2003, Glossary, 18.
71 U.S. Department of Defense, 'FM3–05.301,' 6/24–6/25.
72 U.S. Department of Defense, 'FM3–05.301,' 9/11.
73 Scott Macdonald, *Propaganda and Information Warfare in the Twenty-First Century: Altered Images and Deception Operations* (London: Routledge, 2007), 33.
74 Linebarger, *Psychological Warfare*, 46.
75 Herbert Romerstein, 'Counterpropaganda: We Can't Win without It,' in: J. Michael Waller (ed.), *Strategic Influence: Public Diplomacy, Counterpropaganda, and Political Warfare* (Washington, DC: The Institute of World Politics Press, 2008), 137.
76 Lucas Graves and Federica Cherubini, 'The Rise of Fact-Checking Sites in Europe,' Reuters Institute for the Study of Journalism 2016, available at: https://reutersinstitute.politics.ox.ac.uk/sites/default/files/research/files/The%2520Rise%2520of%2520Fact-Checking%2520Sites%2520in%2520Europe.pdf, 6.
77 Emily Bell, 'The Fact-Check Industry,' *Columbia Journalism Review*, November 10, 2016, available at: www.cjr.org/special_report/fact-check-industry-twitter.php

78 Michael Schellenberger, 'The Censorship Industrial Complex: U.S. Government Support for Domestic Censorship and Disinformation Campaigns, 2016–2022,' Testimony by Michael Schellenberger to the House Judiciary Select Subcommittee on the Weaponization of the Federal Government, March 9, 2023, available at: https://judiciary.house.gov/sites/evo-subsites/republicans-judiciary.house.gov/files/evo-media-document/shellenberger-testimony.pdf

79 Social engineering is also a hacker term for tricking users into violating security procedures for a contrived reason on behalf of the hacker, such as revealing a password or clicking on a link to a malicious website.

80 Marie D. Jones and Larry Flaxman, *Mind Wars: A History of Mind Control, Surveillance, and Social Engineering by the Government, Media, and Secret Societies* (Pompton Plains, NJ: Career Press, 2015), 140–141.

81 This is a standard argument made by proponents of social engineering and is argued quite well in Richard H. Thaler and Cass Sunstein, *Nudge: The Final Edition* (New York: Penguin, 2021).

82 Stanley Milgram, 'Behavioral Study of Obedience,' *The Journal of Abnormal and Social Psychology* 67, no. 4 (1963): 371–378.

83 Stephen Gibson, *Arguing, Obeying, and Defying: A Rhetorical Perspective on Stanley Milgram's Obedience Experiments* (Cambridge: Cambridge University Press, 2019), 18.

84 When the location of the experiment was moved from the respectable campus of Yale University to shabby offices in Bridgeport, the obedience rate dropped from 65 percent to 48 percent. Charles Helm and Mario Morelli, 'Stanley Milgram and the Obedience Experiment: Authority, Legitimacy, and Human Action,' *Political Theory* 7, no. 3 (1979): 324.

85 Solomon Asch, 'Studies of Independence and Conformity: I. A Minority of One against a Unanimous Majority,' *Psychological Monographs* 70, no. 9 (1956): 1–2.

86 Asch, 'Studies of Independence and Conformity,' 11.

87 Mattias Desmet, *The Psychology of Totalitarianism* (London: Chelsea Green, 2022), 99.

88 Desmet, *The Psychology of Totalitarianism*, 94–95.

89 Desmet, *The Psychology of Totalitarianism*, 97.

90 Leon Festinger, 'Cognitive Dissonance,' *The Scientific American* 207, no. 4 (1962): 95.

91 Eric Beerbohm and Ryan W. Davis, 'Gaslighting Citizens,' *American Journal of Political Science* (2021), https://doi.org/10.1111/ajps.12678, 1.

92 Beerbohm and Davis, 'Gaslighting Citizens,' 2.

93 Jacques Ellul, *Propaganda: The Formation of Men's Attitudes* (New York: Vintage Books, 1985), 13.

94 Ishac Diwan and Irina Vartanova, 'Does Education Indoctrinate?' *International Journal of Educational Development* 78 (2020): 2.

95 Yang Jisheng, *The World Turned Upside Down: A History of the Chinese Cultural Revolution* (New York: Farrar Strauss Giroux, 2016), 84.

96 Jisheng, *The World Turned Upside Down*, 104–105.

97 Jisheng, *The World Turned Upside Down*, 88.

98 The 2003 National Assessment of Adult Literacy indicated that only slightly more than half of all Americans have better than basic skills in reading, writing, and math. Americans also do not know the most basic facts about their history or the U.S. Constitution. Bryan Caplan, *The Case against Education: Why the Education System Is a Waste of Time and Money* (Princeton, NJ: Princeton University Press, 2018), 41–45.

99 Caitlin Tilley, 'IQ Scores in the US Have DROPPED for First Time in Nearly 100 Years, Study Suggests—Is Tech Making Us Dim?' *Daily Mail*, March 9, 2023, available at: www.dailymail.co.uk/health/article-11841169/IQ-scores-dropped-time-nearly-100-years.html

100 The mechanisms and policies with respect to U.S. education have been outlined by Charlotte Iserbyt in her book *The Deliberate Dumbing Down of America* (Ravenna, OH: Conscience Press, 2000).

101 Surveys have consistently shown that Americans know almost nothing about the U.S. Constitution or the American political system. Individuals who do not know what rights they have under the Constitution are not going to miss them when they are taken away or eroded. Chris Cillizza, 'Americans Know Literally Nothing about the Constitution,' CNN, September 13, 2017, available at: www.cnn.com/2017/09/13/politics/poll-constitution/index.html

102 Edward Hunter, *Brainwashing: The Story of Men Who Defied It* (Pickle Partners, 2016), 37.

103 Hunter, *Brainwashing*, 170.

104 Hunter, *Brainwashing*, 169.

105 Timothy Thomas, 'Russia's Reflexive Control Theory and the Military,' *Journal of Slavic Military Studies* 17, no, 2 (2004): 240.

106 Can Kasapoglu, 'Russia's Renewed Military Thinking: Non-Linear Warfare and Reflexive Control,' NATO Research Paper, November 2015, available at: www.ndc.nato.int/news/news.php?icode=877, 2.

107 Thomas, 'Russia's Reflexive Control Theory and the Military,' 241.

108 Thomas, 'Russia's Reflexive Control Theory and the Military,' 242.

109 Kasapoglu, 'Russia's Renewed Military Thinking,' 6.

110 Thaler and Sunstein, *Nudge*, 31.

111 Thaler and Sunstein, *Nudge*, 37.

112 Thaler and Sunstein, *Nudge*, 22.

113 Thaler and Sunstein, *Nudge*, 22.

114 Thaler and Sunstein, *Nudge*, 44–45.

115 Roger McNamee, *Zucked: Waking Up to the Facebook Catastrophe* (London: HarperCollins Publishers, 2019), 253–254.

116 Susan Michie, Maartje van Stralen, and Robert West, 'The Behavior Change Wheel: A New Method for Characterising and Designing Behaviour Change Interventions,' *Implementation Science* 6, no. 42 (2011): 1–11.

117 Michie, van Stralen, and West, 'The Behavior Change Wheel,' 4.

118 Susan Michie, Lou Atkins, and Robert West, *The Behaviour Change Wheel: A Guide to Designing Interventions* (Silverback, 2014), 82.

119 Regulatory capture is a theory that posits that regulatory bodies can be captured by the industries that they are meant to oversee through a revolving door, where former regulators join the industry or vice versa. Some agencies even finance themselves by way of collecting fees from the industry they oversee, which can create conflicts of interest as a failure to approve industry proposals results in less agency income.

120 Thomas Kuhn, *The Structure of Scientific Revolutions* (Chicago, IL: University of Chicago Press, 1996).

121 John P.A. Ioannidis, 'Why Most Published Research Findings Are False,' *PLOS Medicine* 19, no. 8 (2005): e1004085.

122 Monya Baker, '1,500 Scientists Lift the Lid on Reproducibility,' *Nature* 533 (2016): 452–454.

123 Danielle Fanelli, 'How Many Scientists Fabricate and Falsify Results? A Systematic Review and Meta-analysis of Survey Data,' *PLOS One* (2009), https://doi.org/10.1371/journal.pone.0005738

124 Andrew Goliszek, *In the Name of Science: A History of Secret Programs, Medical Research, and Human Experimentation* (New York: St. Martin's Press, 2003), 211.

125 Brandon J. Weichert, *Biohacked! China's Race to Control Life* (New York: Encounter Books, 2023), 114.

126 Bernd Lahno, 'Challenging the Majority Rule in Matters of Truth,' *Erasmus Journal for Philosophy and Economics* 7, no. 2 (2014): 54–72.

127 Diego Rinallo and Suman Basuroy, 'Does Advertising Spending Influence the Media Coverage of the Advertiser?' *Journal of Marketing* 73, no. 6 (2009): 33–46.
128 Ben Weingarten, 'FBI Gaslights America over Twitter Files,' *Newsweek*, December 26, 2022, available at: www.newsweek.com/fbi-gaslights-america-over-twitter-files-opinion-1769352
129 Justin Hart, *Gone Viral: How Covid Drove the World Insane* (Washington, DC: Regnery, 2022), XIV.
130 George F. Will, 'Watch for a Return of the Ignominious Disinformation Governance Board,' *Washington Post*, May 25, 2022.

References

Allen, Patrick D. and Gilbert, Dennis P., 'The Information Sphere Domain Increasing Understanding and Cooperation,' NATO Cooperative Cyber Defence Centre of Excellence (2018).

Argemi, Marc and Fine, Gary Alan, 'Faked News: The Politics of Rumour in British World War II Propaganda,' *Journal of War & Culture Studies* 12, no. 2 (2019): 176–193.

Asch, Solomon, 'Studies of Independence and Conformity: I. A Minority of One against a Unanimous Majority,' *Psychological Monographs* 70, no. 9 (1956): 1–70.

Baker, Monya, '1,500 Scientists Lift the Lid on Reproducibility,' *Nature* 533 (2016): 452–454.

Beerbohm, Eric and Davis, Ryan W., 'Gaslighting Citizens,' *American Journal of Political Science* (2021), https://doi.org/10.1111/ajps.12678

Bell, Emily, 'The Fact-Check Industry,' *Columbia Journalism Review*, November 10, 2016, available at: www.cjr.org/special_report/fact-check-industry-twitter.php

Bernays, Edward, *Propaganda* (New York: Horace Liveright, 1928).

Caplan, Bryan, *The Case against Education: Why the Education System Is a Waste of Time and Money* (Princeton, NJ: Princeton University Press, 2018).

Cillizza, Chris, 'Americans Know Literally Nothing about the Constitution,' CNN, September 13, 2017, available at: www.cnn.com/2017/09/13/politics/poll-constitution/index.html

Clizbe, Kent, *Willing Accomplices: How KGB Covert Influence Agents Created Political Correctness, Obama's Hate America First Political Platform, and Destroyed America* (Ashburn, VA: Andemca, 2011).

Connable, Ben, 'Human Terrain System Is Dead . . . What? Building and Sustaining Military Cultural Competence in the Aftermath of the Human Terrain System,' *Military Review* (January–February 2018): 24–33.

Cranston, Maurice, 'Ideology,' *Encyclopedia Britannica*, available at: www.britannica.com/topic/ideology-society, accessed on May 30, 2023.

DeGroot, Gerard, *A Noble Cause? America and the Vietnam War* (London: Longman, 2000).

Delistraty, Cody D., 'The Psychological Comforts of Storytelling,' *The Atlantic*, November 2, 2014, available at: www.theatlantic.com/health/archive/2014/11/the-psychological-comforts-of-storytelling/381964/

Desmet, Mattias, *The Psychology of Totalitarianism* (London: Chelsea Green, 2022).

Dhami, Manddep K., 'Behavioural Science Support for JTRIG's (Joint Threat Research and Intelligence Group's) Effects and Online HUMINT Operations,' Human Systems Group, Information Management Department, DSTL, March 10, 2011.

Diwan, Ishac and Vartanova, Irina, 'Does Education Indoctrinate?' *International Journal of Educational Development* 78 (2020), available at: www.proquest.com/working-papers/does-education-indoctrinate-effect-on-political/docview/2189170361/se-2

Dmytryshyn, Basyl, 'The SS Division "Galicia": Its Genesis, Training, Deployment,' *Nationalities Papers* 21, no. 2 (1993): 53–73.

Doob, Leonard W., 'Goebbels' Principles of Propaganda,' *The Public Opinion Quarterly* 14, no. 3 (1950): 419–442.

Ellul, Jacques, *Propaganda: The Formation of Men's Attitudes* (New York: Vintage Books, 1985).

Engels, Friedrich and Moore, Samuel (eds.), *Manifesto of the Communist Party. By Karl Marx and Friedrich Engels. Authorized English Translation* (Chicago, IL: Charles H. Kerr, 1888).

Fanelli, Danielle, 'How Many Scientists Fabricate and Falsify Results? A Systematic Review and Meta-analysis of Survey Data,' *PLOS One* (2009), https://doi.org/10.1371/journal.pone.0005738

Festinger, Leon, 'Cognitive Dissonance,' *The Scientific American* 207, no. 4 (1962): 93–106.

Flynn, Michael and Cutler, Boone, *The Citizen's Guide to Fifth Generation Warfare* (Resilient Patriot, 2022).

Galula, David, *Counterinsurgency Warfare: Theory and Practice* (New York: Frederick E. Praeger, 1964).

Gibson, Stephen, *Arguing, Obeying, and Defying: A Rhetorical Perspective on Stanley Milgram's Obedience Experiments* (Cambridge: Cambridge University Press, 2019).

Goliszek, Andrew, *In the Name of Science: A History of Secret Programs, Medical Research, and Human Experimentation* (New York: St. Martin's Press, 2003).

Grace, Brent, 'The War for Robert Taylor,' in Daniel H. Abbott (ed.), *The Handbook of 5GW* (Ann Arbor, MI: Nimble Books, 2010), 136–151.

Graves, Lucas and Cherubini, Federica, 'The Rise of Fact-Checking Sites in Europe,' Reuters Institute for the Study of Journalism, 2016, available at: https://reutersinstitute.politics.ox.ac.uk/sites/default/files/research/files/The%2520Rise%2520of%2520Fact-Checking%2520Sites%2520in%2520Europe.pdf

Hart, Justin, *Gone Viral: How Covid Drove the World Insane* (Washington, DC: Regnery, 2022).

Helm, Charles and Morelli, Mario, 'Stanley Milgram and the Obedience Experiment: Authority, Legitimacy, and Human Action,' *Political Theory* 7, no. 3 (1979): 321–345.

Hunter, Edward, *Brainwashing: The Story of Men Who Defied It* (Pickle Partners, 2016).

Ioannidis, John P.A., 'Why Most Published Research Findings Are False,' *PLOS Medicine* 19, no. 8 (2005): e1004085.

Iserbyt, Charlotte, *The Deliberate Dumbing Down of America* (Ravenna, OH: Conscience Press, 2000).

Jeffries, Stuart, 'Why Marxism Is on the Rise Again,' *The Guardian*, July 4, 2012, available at: www.theguardian.com/world/2012/jul/04/the-return-of-marxism

Jisheng, Yang, *The World Turned Upside Down: A History of the Chinese Cultural Revolution* (New York: Farrar Strauss Giroux, 2016).

Jones, Marie D. and Flaxman, Larry, *Mind Wars: A History of Mind Control, Surveillance, and Social Engineering by the Government, Media, and Secret Societies* (Pompton Plains, NJ: Career Press, 2015).

Jowett, Garth S. and O'Donnell, Victoria, *Propaganda and Persuasion* (Thousand Oaks, CA: Sage, 2012).

Kasapoglu, Can, 'Russia's Renewed Military Thinking: Non-Linear Warfare and Reflexive Control,' NATO Research Paper, November 2015, available at: www.ndc.nato.int/news/news.php?icode=877

Kilcullen, David, *Counterinsurgency* (Oxford: Oxford University Press, 2010).

King, Anthony, 'Urban Insurgency in the Twenty-First Century: Smaller Militaries and Increased Conflict in Cities,' *International Affairs* 98, no. 2 (2022): 609–629.

Kuhn, Thomas, *The Structure of Scientific Revolutions* (Chicago, IL: University of Chicago Press, 1996).

Lahno, Bernd, 'Challenging the Majority Rule in Matters of Truth,' *Erasmus Journal for Philosophy and Economics* 7, no. 2 (2014): 54–72.

Linebarger, Paul M.A., *Psychological Warfare* (Eastford, CT: Martino Fine Books, 2022).

Macdonald, Scott, *Propaganda and Information Warfare in the Twenty-First Century: Altered Images and Deception Operations* (London: Routledge, 2007).

McNamee, Roger, *Zucked: Waking Up to the Facebook Catastrophe* (London: HarperCollins, 2019).

Michie, Susan, Atkins, Lou, and West, Robert, *The Behaviour Change Wheel: A Guide to Designing Interventions* (Silverback, 2014).

Michie, Susan, van Stralen, Maartje, and West, Robert, 'The Behavior Change Wheel: A New Method for Characterising and Designing Behaviour Change Interventions,' *Implementation Science* 6, no. 42 (2011): 1–11.

Milgram, Stanley, 'Behavioral Study of Obedience,' *The Journal of Abnormal and Social Psychology* 67, no. 4 (1963): 371–378.

Millerman, Michael, *Inside 'Putin's Brain': The Political Philosophy of Alexander Dugin* (Millerman School, 2022).

Nagl, John, *Learning to Eat Soup with a Knife: Counterinsurgency Lessons from Malaya and Vietnam* (Chicago, IL: University of Chicago Press, 2005).

Nicolette, David, *The Mongol Warlords: Genghis Khan, Kublai Khan, Hulegu, Tamerlane* (Poole, UK: Firebird, 1990).

Pacepa, Ion Mihai and Rychlak, Ronald, *Disinformation: Former Spy Chief Reveals Secret Strategies for Undermining Freedom, Attacking Religion, and Promoting Terrorism* (Washington, DC: WND Books, 2013).

Pampinella, Stephen, 'The Construction of 5GW,' in Daniel H. Abbott (ed.), *The Handbook of 5GW* (Ann Arbor, MI: Nimble Books, 2010), 47–62.

Paul, Christopher, Clarke, Colin P., Grill, Beth, and Dunigan, Molly, 'Moving Beyond Population-centric vs. Enemy-centric Counterinsurgency,' *Small Wars & Insurgencies* 27, no. 6 (2016): 1019–1042.

Pease, Stephen E., *Psywar: Psychological Warfare in Korea, 1950–1953* (Harrisburg, PA: Stackpole Books, 1992).

Qualter, Terence H., *Propaganda and Psychological Warfare* (Burtyrki Books, 2020).

Rees, L.C., 'The End of the Rainbow: Implications of 5GW for a General Theory of War,' in: Daniel H. Abbott (ed.), *The Handbook of 5GW* (Ann Arbor, MI: Nimble Books, 2010), 20–29.

Reveron, Derek (ed.), *Cyberspace and National Security: Threats, Opportunities, and Power in a Virtual World* (Washington, DC: Georgetown University Press, 2012).

Rid, Thomas, *Active Measures: The Secret History of Disinformation and Political Warfare* (New York: Farrar Strauss Giroux, 2020).

Rinallo, Diego and Basuroy, Suman, 'Does Advertising Spending Influence the Media Coverage of the Advertiser?' *Journal of Marketing* 73, no. 6 (2009): 33–46.

Romerstein, Herbert, 'Counterpropaganda: We Can't Win without It,' in: J. Michael Waller (ed.), *Strategic Influence: Public Diplomacy, Counterpropaganda, and Political Warfare* (Washington, DC: The Institute of World Politics Press, 2008), 137–195.

Schellenberger, Michael, 'The Censorship Industrial Complex: U.S. Government Support for Domestic Censorship and Disinformation Campaigns, 2016–2022,' Testimony by

Michael Schellenberger to the House Judiciary Select Subcommittee on the Weaponization of the Federal Government, March 9, 2023, available at: https://judiciary.house.gov/sites/evo-subsites/republicans-judiciary.house.gov/files/evo-media-document/shellenberger-testimony.pdf

Schramm, Wilbur, Katz, Daniel, Kendall, Willmoore, and Vallance, Thedore, 'The Nature of Psychological Warfare,' Research Analysis Corporation, McLean, VA, Technical Memo, April 23, 1954.

Shultz, Richard and Godson, Roy, *Dezinformatsia: Active Measures in Soviet Strategy* (Washington, DC: Pergamon Brassey's, 1984).

Skousen, W. Cleon, *The Naked Communist* (Salt Lake City, UT: Izzard Ink, 2014).

Smith, Rupert, *The Utility of Force: The Art of War in the Modern World* (London: Penguin, 2005).

Taber, Robert, *The War of the Flea: The Classic Study of Guerrilla Warfare* (Lincoln, NE: Potomac Books, 2002).

Thaler, Richard H. and Sunstein, Cass, *Nudge: The Final Edition* (New York: Penguin, 2021).

Thomas, Timothy, 'Russia's Reflexive Control Theory and the Military,' *Journal of Slavic Military Studies* 17, no. 2 (2004): 237–256.

Tilley, Caitlin, 'IQ Scores in the US Have DROPPED for First Time in Nearly 100 Years, Study Suggests—Is Tech Making Us Dim?' *Daily Mail*, March 9, 2023, available at: www.dailymail.co.uk/health/article-11841169/IQ-scores-dropped-time-nearly-100-years.html

Ucko, David H., 'The Five Fallacies of Clear–Hold–Build: Counterinsurgency, Governance and Development at the Local Level,' *The RUSI Journal* 158, no. 3 (2013): 54–61.

———, 'Whither Counterinsurgency: The Rise and Fall of a Divisive Concept,' in: Paul B. Rich and Isabelle Duyvesteyn (eds.), *The Routledge Handbook of Insurgency and Counterinsurgency* (London: Routledge, 2012), 67–79.

Ullman, Harlan K. and Wade, James P., *Shock and Awe: Achieving Rapid Dominance* (Washington, DC: National Defense University, 1996).

U.S. Department of Defense, 'FM3–05.301: Psychological Operations Tactics, Techniques, and Procedures,' Department of the Army, December 2003, Glossary.

———, 'Department of Defense Strategy for Operating in Cyberspace,' July 2011, available at: https://csrc.nist.gov/CSRC/media/Projects/ISPAB/documents/DOD-Strategy-for-Operating-in-Cyberspace.pdf

U.S. Department of State, 'Active Measures: A Report on the Substance and Process of Anti-U.S. Disinformation and Propaganda Campaigns,' August 1986.

U.S. Special Operations Command, 'Operating in the Human Domain,' Version 1.0, August 3, 2015.

Walker, Mason and Matsa, Katerina Eva, 'News Consumption across Social Media in 2021,' Pew Research Center, September 20, 2021, available at: www.pewresearch.org/journalism/2021/09/20/news-consumption-across-social-media-in-2021/

Weichert, Brandon J., *Biohacked! China's Race to Control Life* (New York: Encounter Books, 2023).

Will, George F., 'Watch for a Return of the Ignominious Disinformation Governance Board,' *Washington Post*, May 25, 2022.

Weingarten, Ben, 'FBI Gaslights America over Twitter Files,' *Newsweek*, December 26, 2022, available at: www.newsweek.com/fbi-gaslights-america-over-twitter-files-opinion-1769352

4   Cultural and Cognitive Warfare

Fifth generation warfare is complex and is therefore very difficult to study and to describe. Fifth generation warfare is typically conducted by coalitions of state and non-state actors that operate with very limited coordination utilizing incentives, swarming, and self-organization. A 5GW opponent may infiltrate and/or subtly manipulate organizations that are used as proxies and are not necessarily witting participants in some sort of conspiracy. Sometimes, elites may leverage state resources for their own purposes, which can even be in opposition to official state policies. Often, 5GW relies on ambiguous and covert activity, which can be disguised as being benign or can simply be denied publicly by the aggressors. Practitioners of 5GW will predictably feign ignorance or might not even be identified at all. Four case studies will be used to illustrate 5GW; two are historical, and two are more contemporary. The first example is Stalin's manipulation of the secret police and the party to consolidate his power. The second example is the Arab Spring, which involved the interplay of NGOs, corporations, and activists to facilitate political change in the Middle East. The third example is the attack on U.S. diplomats carried out by an unknown party using directed energy weapons (DEWs). The fourth example is the manipulation of the COVID-19 crisis by China as a means of covert economic warfare. Obviously, any example of 5GW can be expected to be controversial in terms of the interpretation of events since 5GW is a war of narratives and opposing interpretations.

Stalin's Revolution from Above and the Purges

The Stalinist purges of the 1930s, commonly known as the 'Great Terror,' are among the biggest crimes of the 20th century and may have amounted to genocide.[1] The purges started in 1935 and ended in 1939. The initial phase consisted of preparation where the mechanisms for the purges were put in place. This was followed by three major show trials, which provided a narrative of a large-scale anti-Soviet conspiracy. The third phase was mass repression, which targeted particular ethnic groups. The final act was the purging of the NKVD and, in particular, its leadership. Amazingly, the purges were instigated by just one man (Stalin) in a 'conspiracy from above,' which attacked the party, the state, and Soviet society more broadly, with each of these (unwittingly) cooperating in their own subjugation.[2]

DOI: 10.4324/9781003396963-5

Robert Tucker asked 'how could Stalin work his terrorist will on a nation of 165 million?' Tucker suggested that it was owing to '[c]areful advance planning and organization' and 'the incomprehension in Bolshevik minds of what was happening.'[3] In other words, Stalin's strategy depended on extensive preparation, secrecy, and large-scale deception. In short, the Stalinist purges were a masterpiece of 5GW.

The Preparation Phase of the Purges

There is an ongoing historical debate as to whether the purges were premeditated and the extent to which Stalin was actually in control of the process. Robert Tucker has made a compelling argument that the purges were intentional, extensively planned in advance, and tightly controlled by Stalin himself. He pointed out that, before the Great Terror commenced, Stalin put the legal mechanisms and administrative machinery in place that later carried out the purges, since the party was unlikely to participate in its own transformation and was ill-equipped to handle large-scale arrests.[4] According to Tucker,

> Each republic and regional party committee had its own special sector, directly subordinate to and in communication with the main one. Via Poskrebyshev and his staff, Stalin thus presided over a personal power structure ramified throughout the country. It possessed extensive personnel files, served as the main conduit of confidential information and directives, and handled liaison with the security police. No better instrument for a conspiracy from above could have been devised.[5]

In July 1934 the OGPU (the secret police then incorporated into the NKVD) was also given a special board that could nonjudicially sentence anyone to up to five years of exile and labor camps.[6]

What was needed was an event that could justify the initiation of large-scale purges. The opportunity came with the assassination of Leningrad party official and Politburo member Sergei Kirov, who was shot by Leonid Nikolaev in the Smolny Institute on December 1, 1934. Despite some strong arguments that Stalin had ordered the killing of Kirov, the evidence has remained circumstantial and speculative.[7] Donald Rayfield has argued that 'the simplest explanation seems the best: that Leonid Nikolaev was a demented, aggrieved killer acting on his own, aided only in luck by encountering Kirov when he was unguarded.'[8]

In any case, Stalin immediately took action by issuing a new decree that gave the NKVD special powers in terrorism cases, including the power to carry out death sentences soon after an accelerated investigation and judicial process.[9] The new head of the NKVD, Genrikh Yagoda, and prosecutor Andrei Vyshinsky launched an investigation into the assassination to determine whether there was a larger conspiracy. Following the interrogation of Nikolaev, over a hundred people were arrested and shot.[10] The first major political targets were the party officials Gregori Zinoviev and Lev Kamenev, who Stalin considered to be potential rivals.

The Show Trials

In order to purge the Bolshevik Party, Stalin needed to prove a large-scale conspiracy that involved high-level party members. Tucker argued that, in January 1935, Stalin developed the script of the planned show trial of Zinoviev, Kamenev, and others:

> First, the conspiracy turned into a large-scale high-level affair, involving a 'bloc' between foremost Zinovievists and foremost Trotskyists acting under Trotsky's orders from abroad. Second, this conspiratorial 'united center' was said to have been formed toward the end of 1932. Third, it became a terrorist conspiracy primarily against Stalin . . . Fourth, ties between the conspirators and the Gestapo became part of the story. Finally, the anti-Stalin plot was said to have been motivated by the very success achieved under his leadership.[11]

The purpose of the show trials was to have the accused publicly admit to the charges, however contrived and baseless they were, in order to justify more purges in the aftermath of the 'proven conspiracy' against the state. The first show trial took place in August 1936. Thanks to the careful preparation of the defendants involving mental torture and threats against their families, they all confessed to their (fictional) crimes. These included collaboration with the Nazis in the plot against Kirov. Some even demanded in court to be shot for their treason.[12] Soon after the show trial, Yagoda was removed and replaced with Nikolai Yezhov as head of the NKVD.[13]

The second show trial took place in January 1937 and targeted two other Old Bolsheviks, namely Georgy Pyatakov and Karl Radek, as well as lesser-known party officials. As in the first trial, all evidence against the accused was fabricated, and, predictably, all were found guilty of being part of an anti-Soviet, Trotskyist-center conspiracy. The third and final show trial took place in March 1938, with the most prominent victims being Nikolai Bukharin, Alexei Rykov, and Genrikh Yagoda, who were accused of being part of a rightist and Trotskyite conspiracy. With the end of the third show trial, the mass repressions began, 'which were carried out in a series of discrete campaigns by the secret police and targeted specific groups in the population.'[14]

The Purging of the Soviet Military and Mass Repression

Stalin effectively took over the NKVD in 1937 by purging it and replacing the purged individuals with trusted people who would willingly implement the planned large-scale purges of the Red Army and the population. Over 3,000 NKVD operatives were killed in 1937.[15] Stalin feared that the Red Army could stage a coup against him.[16] In May 1937, Marshal Mikhail Tukhachevsky and seven other commanding officers in the Red Army were arrested, tortured during their interrogation, tried in secret, and subsequently shot immediately after the trial.[17] Their supposed guilt was to have conspired with the Nazis, which justified a further purge of the

Red Army that targeted the officers but also lower ranks. At least 34,000 Soviet officers were dismissed, many of whom were later sent to the Gulag.[18]

Soon after the Tukhachevsky trial, the NKVD started its mass operations under NKVD Order 00447 of July 1937, which took the form of secret trials by troika that could sentence enemies of the people to shooting or the Gulag.[19] Quotas for sentencing were provided on the basis of estimated numbers of anti-Soviet elements in a region. A key assumption was that the Germans had been infiltrating the Soviet Union and were organizing espionage and sabotage, which meant that many people were accused of being foreign spies or saboteurs. A major mechanism for the purges of the general population was the use of denunciation. Soviet citizens were encouraged, if not outright pressured, to report potential enemies of the people to the NKVD.[20] More arrests sparked even more arrests, which was a process that quickly escalated out of control. Citizens were getting picked up by the NKVD in the middle of the night, never to be seen again.[21] Many were tortured until they confessed and also implicated others, who would then be purged as well. Altogether, three-quarters of a million people were summarily executed, and over a million people were sent to the Gulag between winter 1937 and autumn 1938, with many never to return.[22]

The Purging of the NKVD

Before 1935, the party controlled the NKVD, but, between 1935 and 1938 when the NKVD was used to purge the party, the NKVD was temporarily in a more powerful position.[23] After hundreds of thousands had been removed from the party and replaced with cadres, and after the purges conducted by the NKVD had escalated too much, it was time for the party to reassert its control over the secret police in late 1938. In September 1938, the Politburo decided to review all NKVD officers with respect to potential hostile elements in their ranks.[24] Many NKVD officers were dismissed, and they were replaced by party cadres.[25] In November 1938, Yezhov resigned as NKVD chief and was replaced by his deputy, Lavrenti Beria, who went on to purge the NKVD of 346 of Yezhov's associates, who were shot.[26] Yezhov was arrested in April 1939, tried in secret, and shot in February 1940 after confessing to having been an enemy of the people and other capital crimes.[27] Yezhov was chiefly blamed for the excessive scale of the purges, which were referred to as Yezhovschina, in order to deflect from the fact that Stalin was the primary architect of the genocide.

The Impact of the Purges

The relentless propaganda of a large-scale Trotskyite conspiracy triggered a frantic and widening search for 'enemies of the people' who had to be eliminated. Owing to mass psychosis, a climate of constant fear, and various individual incentives such as promotion and privileges, a large number of people actively cooperated with Stalin's nefarious plot, even though it often led to their own demise. Interestingly, many Soviets were confused as to why the purges were happening and who

Table 4.1 The Stalinist Purges as 5GW

Domain	Human domain, Soviet society
Adversaries	Stalin, the NKVD
Objectives	Eliminate any threats to Stalin's rule, transform the Soviet system, consolidate power
Violence	Dispersed, targeted at particular individuals and groups
Outcome	Key institutions were purged, and their organizational cultures were transformed

Source: Created by author

authorized them, speculating that Stalin was not aware of them and that they were masterminded by 'mysterious anti-Communist enemies in high places,' as many refused to believe that they were Stalin's doing.[28]

Stalin first used the NKVD to purge the party and then used the transformed party to reassert control and purge the NKVD. Key institutions such as the Bolshevik Party, the Red Army, and the NKVD were transformed by purging the leadership and thereby permanently changing the organizational culture. Mark Safranski argued that it was common practice

> to liquidate not only the holder of an important post in an organization, but his immediate replacement as well (and not infrequently, the replacement's replacement), thus not only atomizing existing social networks, but terminating institutional memory in the bargain as the documentary evidence was purged with the same severity as the staff.[29]

Safranski also pointed out that the mass purges from 1937 to 1938 were not randomly removing people but were targeted at particular groups such as the Polish Communist Party in exile, the Ukrainian Communist Party, and the Georgian Communist Party, which were considered by Stalin to be a threat owing to their nationalism. Although the Stalinist purges were atypically violent for a case of 5GW, they relied primarily on the manipulation of proxies and psychological terror to carry out a revolution from above.

The Arab Spring

The Arab Spring was a wave of protests and revolutions across the Middle East and North Africa that started in late 2010 in Tunisia; it resulted in the overthrow of the governments of Tunisia, Egypt, Yemen, and Libya and facilitated political reforms in Jordan, Oman, and Morocco. The main characteristic of the revolutions was the use of social media by local activists to spread news about government abuses and organize a series of mass protests. A key argument made at the time was that digital technologies were liberating as information could no longer be controlled by tyrants in the internet age.[30] After only a month of protests, the Ben Ali government in Tunisia fell, and, after just 18 days of protest, the Mubarak government in Egypt fell. The popular perception of the Arab Spring was that of an organic event that caught the U.S. government completely by surprise.[31] Although the people of the Middle East had plenty of reasons to revolt against their corrupt, tyrannical governments, it is important to also consider the role of outside state actors and NGOs that prepared the ground for the uprisings.

American Democracy Promotion

The U.S. government has actively promoted democratization worldwide since 1941 when it established Freedom House.[32] Although Freedom House presents itself as an NGO that is independent from the U.S. government, the connections are fairly obvious, as over 90 percent of its funding comes from federal grants.[33] In essence, Freedom House produces and disseminates propaganda in furtherance of U.S. soft power. Freedom House has also provided cover for the CIA, and there have been many connections to the intelligence community (IC).[34]

It was Gene Sharp who, in some sense, 'weaponized' democracy promotion beyond propaganda by developing nonviolent strategies for resisting tyranny in the 1960s and 1970s.[35] Sharp founded the Albert Einstein Institute (AEI), which promotes nonviolent strategies for democratization, in 1983. In the same year, the Reagan administration created the National Endowment for Democracy (NED). NED is almost completely funded by the U.S. State Department but claims to be independent from the U.S. government. *Washington Post* journalist David Ignatius has called it 'the sugar daddy of overt operations, a quasi-private group headed by Carl Gershman that is funded by the U.S. Congress.'[36]

The NED works closely together with the National Democratic Institute (NDI) and the International Republican Institute (IRI), which were also founded in 1983. Critics claim that the NED is little more than an extension of the State Department and the CIA. William Blum quoted one of the founders of the NED, Allen Weinstein, who said in 1991: 'A lot of what we do today was done covertly 25 years ago by the CIA.'[37] Similarly, CIA whistleblower Phil Agee claimed that the NED was simply a CIA 'sidekick' used by the agency to funnel money to opposition groups through affiliated foundations.[38] Most notably, the NED also played some role in ending communist rule in Poland by providing some of the funds given to Solidarity ($9 million) through the international labor union AFL-CIO (American Federation of Labor and Congress of Industrial Organizations).[39]

U.S. objectives in the Middle East were, from the 1990s, to promote democracy as part of a 'democratic expansion' grand strategy of the Clinton administration. The U.S. State Department announced the Middle East Partnership Initiative (MEPI) in 2002, which made $90 million available for projects that support democratization.[40] One of the objectives behind the decision of the George W. Bush administration to invade Iraq was to 'sow seeds of democracy and peace in the Middle East.'[41] He stated in a speech at the NED that '[m]any Middle Eastern governments now understand that military dictatorship and military rule are a straight, smooth highway to nowhere.'[42] In 2004, his administration launched the Greater Middle East Initiative (GMEI) with the goal of 'broadly supporting economic and human development, democracy, and the expansion of civil society' so that the Middle East would no longer be a breeding ground for tyranny and terrorism.[43]

President Obama's Middle East policy regarding the goal of democratization remained unchanged from his predecessor's. The *New York Times* reported that 'President Obama ordered his advisors last August [2010] to produce a secret report on unrest in the Arab world, which concluded that without sweeping political changes, countries from Bahrain to Yemen were ripe for popular revolt.'[44] The Presidential Study Directive 11 identified flashpoints such as Egypt and underlying causes of instability such as youth bulges, weak education systems, poor economic performance, and the proliferation of social media.[45] In short, at the outset of Obama's presidency, there was an opportunity to remake the Middle East by facilitating revolution in order to move it in a favorable direction for the United States.

The U.S. State Department and the Training of Activists

The State Department provided funding, through a network of NGOs, to Arab activists and political groups. For example, USAID spent on average $20 million per year on democratization in Egypt from the late 1990s.[46] More money was funneled through the NED and a couple of other pro-democracy NGOs, including Freedom House, NDI, AEI, and the Center for Applied Non-Violent Actions and Strategies (CANVAS).[47] All of these pro-democracy NGOs receive a large portion, if not all, of their funding from the U.S. government. They then disperse the money, through grants and workshops, to groups and individuals that the U.S. government could not sponsor directly. NED documents showed that it spent $120,000 on funding an Egyptian police officer who used social media to incite violence in his native country while claiming refugee status in the United States.[48] Another NED grant, of $318,757, was awarded to Egyptian labor organizations to 'advocate for and defend worker rights, strengthen respect for the rule of law, and build bridges between Egyptians and other labor movements' after the start of the April 6 Movement that staged a strike that turned violent in 2008.[49]

In 2009, the State Department provided $57 million, for a period of three years, to fund a program that taught foreign dissidents the art of evading government surveillance (encryption, use of firewalls, anonymization, etc.). According to *Time* magazine, this program trained 'more than 10,000 bloggers, journalists and activists . . . in 10 languages through 50 programs, and hundreds of thousands more have accessed materials and guides published by the groups.'[50] The *New York Times* reported that

[s]ome Egyptian youth leaders attended a 2008 technology meeting in New York, where they were taught to use social networking and mobile technologies to promote democracy. Among those sponsoring the meeting were Facebook, Google, MTV, Columbia Law School and the State Department.[51]

Freedom House organized similar workshops for 16 Egyptian activists in 2009, and CANVAS trained Egyptian activists in Belgrade in the same year.[52]

The Digital Uprisings in Tunisia and Egypt

The Arab Spring was reportedly triggered by WikiLeaks releasing confidential U.S. State Department cables that discussed the corruption in various Middle Eastern countries. The revelations were widely discussed in the international media, which meant that 'the flames of revolt were stoked, industriously and ceaselessly, by the media, courtesy of what it was learning by sifting through piles of documents amassed by WikiLeaks.'[53] Interestingly, WikiLeaks may have started out as an intelligence operation, with founding members having connections to the NED and various U.S. foundations.[54] The self-immolation of the street vendor Mohamed Bouazizi in front of a municipal office on December 17 sparked further outrage. The images of the burned Bouazizi ended up on Facebook and were used effectively by activists to mobilize the public against the state.[55]

Mass protests were organized using social media, and, later, social media reports clearly contradicted the government's propaganda that tried to downplay the events. With mounting dissent, the Ben Ali government responded by blocking Facebook and other social media platforms and arresting activists and bloggers. The regime even made an attempt to identify activists using Facebook by systematically stealing login information, which was thwarted by Facebook itself.[56] After a mass protest in Tunis on January 14, 2011, in which more than 100,000 Tunisians participated, Ben Ali stepped down.

In Egypt, Google executives became actively involved in regime change. One of the key players in the events in Egypt in early 2011 was the Google marketing manager Wael Ghonim, who helped ignite the uprising.[57] He was officially on leave from Google when he traveled to Egypt to set up a Facebook group/site named 'Kullena Khaled Saeed' ('we are Khaled Saeed'). Khaled Saeed was a 28-year-old protester who was killed by Egyptian security forces. Ghonim demanded accountability from the Egyptian government and incited widespread protests leading up to the mass protest on Tahrir Square on January 25, 2011. At one point, the Facebook page had 800,000 members.[21] Ghonim was visited by Jared Cohen during the revolution, and the meeting occurred just one hour before Ghonim was arrested by the Egyptian police.[58]

Agence France-Presse (AFP) reported that, at the height of the Arab Spring, in February 2011, the State Department organized a training session for bloggers and activists from Tunisia, Egypt, Syria, and Lebanon. They would go back home and teach other activists, which would then produce a ripple effect.[59] Google and Facebook also aided activists in Tunisia and Egypt by denying their respective governments access to user data, something they handle more liberally in many other countries, including the United States. Although Facebook temporarily removed the popular pages of the two main activist groups (Kullena Khaled Saeed and April 6 Movement) in November 2010, it not only restored the services but actively protected the identities of the activists who were organizing the protests. Google provided technological assistance to activists by creating a workaround that enabled them to send Twitter messages over telephone lines, using voice recognition that transcribed speech and tweeted it, after the Mubarak government shut down the internet on January 28, 2011.[60] Jared Cohen pointed out in an interview that closing down the internet had exactly the opposite effect of what was intended. People hungry for information turned to the streets. As a result, the protests became much bigger and forced Mubarak out.[61]

Jared Cohen later wrote a book with Eric Schmidt, the Google CEO, on *The New Digital Age*, which discussed digital revolutions and the Arab Spring.[62] Billionaire Eric Schmidt was a frequent visitor to the Obama White House. Google staffers visited the White House 427 times between 2009 and 2015, indicating the very close relationship between the Obama administration and Google.[63] It is not a stretch to suggest that Ghonim coordinated his activities with Google, and that Google coordinated its role in the Arab Spring with the White House, each of them acting somewhat autonomously and perhaps being motivated by different goals.

The Impact of the Arab Spring

The revolutions in Tunisia and Egypt were supported by U.S. public diplomacy and other diplomatic measures. Leaked State Department cables indicate that the U.S. embassy in Cairo was in constant contact with activists in the years leading up to the revolution while also pressuring the Egyptian authorities to release dissidents detained by the police.[64] Secretary of State Hillary Clinton issued a clear warning to Mubarak on January 28, 2011, by saying that '[w]e urge the Egyptian authorities to allow peaceful protests and to reverse the unprecedented steps it has taken to cut off communications,' effectively undermining the efforts of the authorities to suppress the protests.[65] On February 1, 2011, President Barack Obama demanded, during a telephone call, that President Mubarak immediately step down, which he reluctantly did more than a week later.[66] Although the U.S. media were very upbeat about the Arab Spring in 2011, it soon became obvious that the U.S. attempt to democratize the Middle East had failed, not just in Egypt but elsewhere as well. The most problematic consequence was the descent of Libya and Syria into civil war and long-lasting chaos.

Particularly interesting is the Russian perception of the Arab Spring, as Russia treated the wave of revolutions as a new mode of conflict. General Gerasimov penned an article in 2013 that contained observations on the Arab Spring. He wrote:

> The experience of military conflicts—including those connected with the so-called color revolutions in North Africa and the Middle East—confirms that a perfectly thriving state can, in a matter of months and even days, be transformed into an arena of fierce armed conflict, become a victim of foreign intervention, and sink into a web of chaos, humanitarian catastrophe, and civil war.[67]

Table 4.2 The Arab Spring as 5GW

Domain	Human domain, society
Adversaries	U.S. government, NGOs, corporations, local activist groups
Objectives	Replace Middle Eastern autocracies with moderate Islamic democracies
Violence	Nonviolent
Outcome	Two regimes overthrown

Source: Created by author

The Russian government held its Third International Security Conference in Moscow in 2014, where top Russian and some international government officials and military officers discussed how color revolutions are employed as an aggressive strategy to overthrow hostile governments.[68] Putin stated in the same year that '[i]n the modern world extremism is being used as a geopolitical instrument and for remaking spheres of influence. We see what tragic consequences the wave of so-called color revolutions led to.'[69] Not surprisingly, the United States-funded pro-democracy NGOs and their employees are no longer welcome in a growing number of countries.[70]

Havana Syndrome

The term Havana Syndrome relates to mysterious 'health incidents' that have affected U.S. diplomats and intelligence officers, starting with a number of cases that occurred in late 2016 and that were first reported by the media in summer 2017.[71] Six of those affected had to be flown from Cuba to Miami for emergency medical treatment. At least 21 American diplomats and some Canadian diplomats had suffered symptoms by 2017.[72] The initial theory was that the diplomats had been subjected to an attack using an unknown type of acoustic weapon because the victims reported chirping noises when they suffered the incidents.[73] Some of the symptoms included headaches, dizziness, insomnia, ear pain, temporary loss of hearing, weakened sense of balance, diminished memory, inability to regulate emotion, and indications of traumatic brain injury and concussion.[74] In December 2017, senior CIA officer Marc Polymeropoulos experienced a sudden onset of nausea in his Moscow hotel room, suffering a second attack two days later in a Moscow restaurant. He had to be flown out of the country and developed permanent symptoms similar to those of the diplomats in Havana, which forced him to retire from his CIA career of 26 years in 2019.[75]

Several diplomats in Guangzhou, China, suffered from similar symptoms, including brain injury, in spring 2018.[76] It started with one embassy employee hearing sounds and getting sick in late 2017, with more cases appearing in April 2018. Mark Lenzi, a security engineer at the consulate, had to be flown out of China together with his family.[77] Further incidents involving U.S. and Canadian government personnel (diplomats, intelligence officers, and staffers) were reported in several locations around the world, including Cuba, China, Russia, Austria, India, Vietnam, Colombia, Uzbekistan, Kirgizstan, and Washington, DC.[78] Suspected attacks occurred in a variety of settings such as 'residencies, on the street, in vehicles, and even at U.S. secure facilities.'[79] The incidents have seemingly escalated in geographic scope, frequency, and severity since late 2016.[80] Altogether, over 1,000 people had been diagnosed with Havana Syndrome as of February 2022.[81]

The Investigations

The State Department conducted an investigation and arranged for the medical examination of the affected diplomats to determine the cause of the symptoms. The University of Pennsylvania Center for Brain Injury and Repair found a number of conditions that could be objectively tested, such as evidence of concussion in persons with no history of brain injury and temporary auditory impairment.[82] With the help of neuroimaging, it was found that there were significant 'differences in whole brain white matter volume, regional gray and white matter volume, cerebellar tissue microstructural integrity, and functional connectivity in the auditory and visuospatial subnetworks but not in the executive control subnetwork' compared with a control group.[83]

Some of the affected diplomats have sought legal remedies to get compensation from the government. Mr. Zaid, who is a lawyer representing some of them, stated: '"It's sort of naïve to think this just started now" . . . Globally, he added, covert strikes with the potent beams appear to have been going on for decades.'[84] Furthermore, the National Security Agency gave Zaid 'a statement on how a foreign power built a weapon "designed to bathe a target's living quarters in microwaves, causing numerous physical effects, including a damaged nervous system."'[85]

The U.S. State Department tasked the National Academy of Sciences (NAS) to advise on how U.S. diplomats could be protected. A standing committee of 19 scientists led by David Relman, a professor of microbiology at Stanford University, reviewed and studied Havana Syndrome and produced a final report that was declassified in December 2020.[86] The committee published its findings in December 2020. The committee considered delusions and hallucinations that may have been caused by stress but concluded that '[t]he acute initial, sudden-onset, distinct and unusual symptoms and signs described in some affected DOS personnel (see Section 3 and CDC Report) cannot be ascribed to psychological and social factors in the absence of patient-level data.'[87] It suggested that 'many of the cognitive, vestibular, and auditory effects observed in DOS personnel are most consistent with modulated, or pulsed, RF biological effects,' citing Soviet/Russian research on the effects of pulsed and continuous RF energy on humans.[88] Referring to other studies in the field, the report stated: 'Pulsed RF effects on the nervous system can include changes to cognitive . . . behavioral . . . vestibular . . . EEG during sleep . . . and auditory . . . function in animals and humans.'[89]

Congress subsequently awarded long-term emergency health benefits to the affected diplomats, thereby acknowledging that their ordeal was serious, significant, real, and not imagined.[90] Affected diplomats have been awarded lump-sum compensation of between $140,000 and $187,000 for hardship, permanent injury received on duty, and subsequent loss of employment.[91] More details are contained in Bill H.R.4914 that the House of Representatives passed on August 3, 2021, authorizing the compensation. The bill states as its rationale: 'To impose sanctions against foreign persons and foreign governments in response to certain clandestine attacks on United States personnel, and for other purposes.'[92] In particular, it suggested, 'United States personnel have suffered persistent brain injuries after being targeted in attacks that have been increasing in number, geographic location, and audacity.'[93] The covert attacks would be 'continuing and [the] expanding scope of these attacks has become a serious security concern that is also undermining the morale of United States personnel, especially those posted at overseas diplomatic missions.'[94]

Indications for Deliberate Attacks with DEWs

An IC expert panel report on Havana Syndrome concluded that '[t]he signs and symptoms of the AHI are genuine and compelling,' 'that they cannot be explained by known environmental and medical conditions,' and that '[e]lectromagnetic energy, particularly pulsed signals in the radiofrequency range, plausibly explain core characteristics.'[95] Although many experts agree that a DEW attack can produce the effects observed in Havana Syndrome victims, the IC as a collective claimed that there was no evidence for a 'global campaign to harm or collect intelligence on US intelligence personnel.'[96] This claim was again made in an 'Updated Assessment of Anomalous Health Incidents' released as a summary in March 2023, which stated '[m]ost IC agencies judge it is very unlikely a foreign adversary played a role,' and that there was 'no credible evidence that a foreign adversary has a weapon or a collection system that is causing AHIs.'[97] Case closed? Hardly.

First of all, the apparent contradictions between earlier investigations and the most recent conclusions are glaring and raise more questions than they answer. Some experts clearly believe that DEWs could produce the effects observed in Havana Syndrome victims, but the IC cannot find evidence that this actually happened, declaring a covert attack therefore to be a 'very unlikely' cause of Havana Syndrome.[98] This leaves two possibilities: Either Havana Syndrome does not exist, or the IC has been unable to determine its true cause owing to insufficient data or poor analysis. It has been argued that the IC, by 'fixating on finding a common thread between all Havana Syndrome reports rather than examining the most credible and concerning reports better supported by intelligence reporting, the intelligence community has given itself analytical space to dispute all reports.'[99] The inability to find evidence could be due to 'the intelligence community's fear of the foreign policy ramifications that would result from finding suspected Russian culpability.'[100]

There is a real dilemma for an attacked government in how to deal with a covert attack and in what kind of narrative to frame it for the public. David Ignatius stated that

Table 4.3 The Havana Syndrome as 5GW

Domain	Human domain, perception, brain
Adversaries	Unknown
Objectives	Create uncertainty and fear, deception
Violence	Violent but highly targeted and dispersed
Outcome	Minor disruption of U.S. foreign policy, strategic uncertainty

Source: Created by author

> [t]hese mysterious attacks are a policymaker's nightmare. You can't accuse another country of warlike assaults without solid facts; the Iraqi WMD fiasco taught a generation of intelligence analysts that lesson. But if you don't hold rogue actors accountable, how do you deter future attacks?[101]

If a government acknowledges the attacks, it looks weak and would be under pressure to do something, but what can it do? If it denies the attacks, seemingly knowing that they are real, it also looks weak and, on top of that, like a dishonest hypocrite, which undermines morale within the government.

What an adversary may want to achieve by attacking low-level diplomats and other government personnel is the creation of uncertainty, fear, and a feeling of helplessness in U.S. decision-makers. Havana Syndrome allegedly '"dramatically hurt" morale in the diplomatic service and affected recruitment' for the foreign service.[102] There are likely State Department and IC officials and even members of Congress who are now afraid to become the next victim of covert attacks that cannot be proven and from which they cannot be protected.

The Manipulation of the COVID-19 Crisis

SARS-CoV-2, the virus that causes the disease named COVID-19, was apparently discovered on December 27, 2019, after an elderly couple had fallen ill in Wuhan, China.[103] There may have been COVID-19 cases as early as August 2019.[104] The outbreak was formally announced by the Chinese government on December 31, 2019.[105] Initially, the Huanan seafood market was identified as the source of the outbreak, with speculation by Chinese scientists that infected bats were consumed there, resulting in a transmission from bats to humans.[106] On January 31, 2020, the WHO declared the outbreak a public emergency of international concern. Cases began to be reported worldwide, with an unusually high number of cases in Northern Italy. The WHO declared a pandemic on March 11, 2020, and soon thereafter, on March 13, 2020, the Trump administration declared a national emergency, with many states implementing a lockdown on March 16, 2020. The question of the true origin of the virus and the possibility of a lab leak from the Wuhan Institute of Virology (WIV) emerged early on, and it became a partisan issue. The issue became so extremely politicized that hardly any rational or objective debate was possible in 2020 and 2021.[107]

The COVID-19 Origin Debate

In 2020, the mainstream U.S. media heavily promoted the idea of a natural origin of COVID-19 and even attacked any suggestion that it came out of a lab as an expression of racism or a conspiracy theory.[108] However, there are many reasons to believe that a lab leak is a probable, if not the most probable, cause. First of all, some researchers, such as Jeffrey Sachs, the Lancet chair of the COVID-19 investigation, have pointed out that nobody has been able to find the animal that supposedly first transmitted the virus to humans, which means that the natural origin hypothesis is essentially unsupported.[109]

Second, there is plenty of circumstantial evidence for a lab leak. The WIV had conducted research related to SARS and bats in the lab of Dr. Shi Zhengli, and there is also a history of SARS lab leaks that occurred in other countries prior to the SARS-CoV-2 outbreak.[110] A fact sheet published by the State Department of the outgoing Trump administration on January 15, 2021, indicated that 'several researchers inside the WIV became sick in autumn 2019,' that 'RaTG13, the bat coronavirus [was] identified by the WIV in January 2020 as its closest sample to SARS-CoV-2 (96.2% similar),' and that '[t]he WIV has a published record of conducting "gain-of-function" research to engineer chimeric viruses.'[111]

A U.S. Senate report claimed that '[t]he preponderance of information supports the plausibility of an unintentional research-related incident that likely resulted from failures of biosafety containment during SARS-CoV-2 vaccine-related research.'[112] Former CDC Director Robert Redfield, White House COVID-19 coordinator Deborah Birx, and FBI Director Christopher Wray publicly stated that they considered the lab leak theory very likely or very plausible.[113]

If SARS-CoV-2 was developed in a lab, it was very likely developed as a bioweapon based on 'gain of function' research that makes a virus more transmissible, as claimed by Chinese virologist Dr. Li-Meng Yan.[114] She suggested that '[a]lthough it is not easy for the public to accept SARS-CoV-2 as a bioweapon due to its relatively low lethality, this virus indeed meets the criteria of a bioweapon.'[115] The State Department fact sheet also pointed out that '[f]or many years the United States has publicly raised concerns about China's past biological weapons work,' and that

> [d]espite the WIV presenting itself as a civilian institution, the United States has determined that the WIV has collaborated on publications and secret projects with China's military. The WIV has engaged in classified research, including laboratory animal experiments, on behalf of the Chinese military since at least 2017.[116]

Information on Chinese biological warfare research is spotty owing to the extreme secrecy inherent in such research programs. An earlier congressional report pointed out that France had suspected China of having a biological warfare program after the first SARS outbreak in 2003, and that France was therefore reluctant to provide the technology for a Biosafety Level 4 lab in China.[117] The lab was eventually built by Chinese and French firms for a cost of $44 million and completed in 2014.[118] The PLA has shown an interest in biological warfare in its strategic writing. A book by Professor Ji-Wei Guo titled *The War of Life-making Right of Reconstruction of Military Strategy*, published in 2010, openly advocated offensive biowarfare.[119] It is, hence, likely that SARS-CoV-2 was developed in the WIV, and that its development was possibly connected to an illegal biowarfare research program. Whether the virus leaked accidentally or was released on purpose may forever remain a mystery, but there is no doubt that China sought to manipulate the perception of the COVID-19 crisis once it started.

The Cover-Up

It is a reasonable assumption that, if China was responsible for a lab leak of a virus possibly associated with biowarfare research, the Chinese government would seek to hide evidence, frustrate any serious investigations into the incident, and put out disinformation to misdirect and distract the world from what had really happened. First of all, the Chinese government delayed alerting the world to the outbreak and also, initially, took few measures to contain the outbreak. On January 20, 2020, a WHO team visited Wuhan to get more information from Chinese scientists about the outbreak, but they downplayed the incident while not sharing much information.[120] As the crisis unfolded, the WHO undertook further efforts to obtain raw data about the virus such as blood samples from December 2019, but Chinese scientists were unwilling to provide them.[121] Senator Michael McCaul's report indicated that the cover-up may have started as early as September 2019, stating '[t]he WIV's sequence library was taken offline in September 2019 and is not "accessible for people to check."'[122] The clear lack of evidence for a natural origin did not stop the WHO from dismissing the lab leak hypothesis.[123]

Not only was a real investigation made impossible by access to the WIV and research data being denied, but there was a clear effort to deceive the world. According to McCaul,

> the Chinese Communist Party (CCP) and the World Health Organization (WHO) went to great lengths to cover up the initial epidemic, and how their cover-up likely turned what could have been a local outbreak into a global pandemic. The CCP detained doctors in order to silence them, and disappeared journalists who attempted to expose the truth. They destroyed lab samples, and hid the fact there was clear evidence of human-to-human transmission. And they still refuse to allow a real investigation into the origins. At the same time, the WHO, under Director General Tedros, failed to warn the world of the impending pandemic. Instead, he parroted CCP talking points, acting as a puppet of General Secretary Xi.

The bottom line is that it is very likely that the Chinese government was aware of the outbreak as early as the second week of December 2019 and did nothing to alert the world or to contain the spread of the virus from China to other countries.[124] This is, at a minimum, negligence and, at a maximum, an attempt to exploit a global spread of the virus for political and economic gain.

Chinese COVID-19 Propaganda

There are many indications that, from January 2020, the CCP started an aggressive online campaign to shape the perception of the outbreak. A computerized content analysis of 134,000 Chinese government and media tweets suggested that 'China crafted an intricate campaign clearing its name' by first downplaying the crisis, then denying that the virus came from China, and finally attacking the accuser.[125] Most notably, in March 2020, a spokesman for the Chinese Ministry of Foreign Affairs even tweeted that the U.S. military had brought coronavirus to China and called it the 'USA virus.'[126]

More nefarious were the English-language tweets by Chinese trolls and bots that targeted Western audiences, which were boosted by inauthentic views, likes, shares, tweets, and retweets through fake accounts.[127] The *New York Times* reported:

> United States intelligence agencies have assessed that Chinese operatives helped push the messages across platforms, according to six American officials, who spoke on the condition of anonymity to publicly discuss intelligence matters. The amplification techniques are alarming to officials because the disinformation showed up as texts on many Americans' cellphones, a tactic that several of the officials said they had not seen before.[128]

An article in *Tablet Magazine* found evidence of social media manipulation and disinformation. It suggested that

> [i]nternational COVID-19 hysteria began around Jan. 23, when 'leaked' videos from Wuhan began flooding international social media sites including Facebook, Twitter, and YouTube—all of which are blocked in China— allegedly showing the horrors of Wuhan's epidemic and the seriousness of its lockdown.[129]

The videos showed dramatic scenes of people collapsing on the streets, people getting locked down in Wuhan by having their front doors welded shut, and scenes of crowded hospitals—all of which may have been carefully staged to make the virus appear to be much more dangerous than it was. Nowhere else in the world did COVID-19 cause people to collapse on the streets, which makes these early videos particularly suspicious.

Over 250,000 inauthentic social media accounts were spreading coronavirus disinformation that severely exaggerated the threat, first targeting Italy in a concerted effort to manipulate Italians into accepting extreme measures. The Chinese trolls flooded Twitter with messages that attacked any head of state or governor who refused to lock down their state, likening their refusal to genocide and an attack on human rights.[130] Graphika discovered the Chinese Spamouflage Dragon network, which sends political spam messages through social media with inauthentic AI-generated profile pictures and in the English language. The messages praised the Chinese Communist Party for its handling of the COVID-19 crisis while blaming the United States and President Trump for the outbreak and the riots in the United States.[131] Many of the propaganda videos posted on YouTube had an election theme where President Trump's policies towards China and Chinese companies were attacked, which likely hurt Trump's election campaign.

U.S. media and the science community simply went along with China's narrative of the pandemic owing to Chinese co-optation and the massive conflicts of interests in the U.S. scientific community.[132] In one incident, the *New York Times* published a heavily biased opinion editorial by a Chinese scientist, Yi Rao, who lamented that the U.S. government had failed to learn from the example of how China dealt with the pandemic, a shortcoming that allegedly killed his uncle in New York while his family in Wuhan was safe.[133] As it turned out, Rao is not only a proud member of the CCP but is also the founder of the Chinese Thousand Talents program, which has been used for recruiting spies and conducting scientific espionage.[134]

Impact of the Pandemic

The COVID-19 crisis is the most consequential international crisis of the 21st century so far, not only because of the humanitarian impact, but more significantly because of the economic and political impacts. As of May 2023, there had been 6.7 million COVID-19 deaths reported worldwide, of which 1.16 million deaths, or 17 percent of the total, occurred in the United States, despite the U.S. population being only 4 percent of the world population.[135] The pandemic contributed greatly to a spike in mental health diagnoses, with an estimated 39 percent of Americans suffering from anxiety and 32 percent suffering from depression in 2021.[136]

The United States was also disproportionately impacted economically. The federal government had provided a total of $4.6 trillion in COVID-19 relief funding as of January 2023.[137] Government debt increased from $26 trillion in mid-2020 to almost $32 trillion in mid-2023. According to research by the University of California, the direct and indirect costs of the pandemic to the United States were estimated to reach $14 trillion by the end of 2023, which would be 20 times more than the economic costs of the 9/11 attacks and 4 times more than any other national disaster.[138] The U.S. government is now on an unsustainable fiscal path.[139] Owing to rising bond yields, the estimated annualized cost of servicing the national debt exceeded $800 billion in March 2023, which is higher than the U.S. defense budget.[140]

The pandemic resulted in unprecedented political measures, based on the Chinese model, that impacted the rights and freedoms of Americans, such as lockdowns with mandatory business closures and travel/movement restrictions, mask mandates, social distancing, mandatory COVID-19 testing, and mandatory COVID-19 vaccinations.[141] Supreme Court Judge Neil Gorsuch said that the COVID-19

Table 4.4 Manipulation of the COVID-19 Crisis as 5GW

Domain	Human domain, (world) society
Adversaries	China, WHO, civil society actors
Objectives	Inflict economic damage and cause political instability in the United States
Violence	Violence was very dispersed and appeared unintended
Outcome	Chinese economic and political objectives were largely achieved

Source: Created by author

measures were 'among the greatest intrusions on civil liberties in the history of the nation.'[142] Richard Fleming argued that

> [t]he best weapon doesn't kill people, it devastates and demoralizes them . . . The best weapon to devastate a country is one that removes the will of the people to fight. It effectively diminishes the lifestyles of the enemy, reducing the security of life as the enemy knows it and replaces that security and freedom with fear and uncertainty. SARS-CoV-2 has done exactly that. It has devastated economics, removed the personal freedoms people were used to, reduced goods and services, and turned friends against friends and family members. It has divided nations and people.[143]

It appears that the COVID-19 pandemic was perhaps a 'proof-of-concept' test by the CCP to see how much damage could be done by the release of a bioweapon, which could suggest that another, perhaps much more deadly, pandemic can be expected.[144] However, even if a second pandemic does not materialize, the damage done by the last one is already catastrophic.

Conclusion

Out of the four examples of 5GW discussed, the Stalinist purges are the most straightforward case of 5GW, as a tiny power elite was able to manipulate various state and party organizations for its own personal benefits. Stalin was waging war on the Bolshevik Party, the Red Army, regional party organizations, and various societal groups he deemed to be a threat. He could not do it without the skillful manipulation of the party, the secret police, and the Soviet public at large, which remained oblivious as to who instigated the purges and why they happened. The Arab Spring example is more controversial, as the U.S. government denied any responsibility and as Russian thought on the matter has been characterized as paranoid. The reality remains that the U.S. State Department and affiliated NGOs had trained and funded Middle Eastern activists and political groups for years, which played a key role in the events of late 2010 and early 2011. Whether this can be called a new mode of warfare is merely a matter of perspective. Havana Syndrome has been well documented, and some experts strongly suggest that it is likely caused by DEW attacks, although it remains hard to prove it conclusively. For the U.S. government, there is a real dilemma to acknowledging these covert attacks as they can be considered acts of war and as it is hard to accuse any government without strong evidence. If Havana Syndrome is caused by DEWs, it is a perfect example of 5GW—a covert attack that is hard to prove and even harder to attribute, causing uncertainty and fear among U.S. officials. The last example is by far the most controversial owing to the global scale and the extremely high stakes. Regardless of whether or not SARS-CoV-2 came out of a lab, it would be hard to deny the fact that the Chinese government manipulated the perception of the COVID-19 crisis, trying to deflect from its own responsibility. If China's government had acted decisively in late summer 2019, the outbreak might have been contained at an early stage. Instead, it opted to cover up the outbreak, then exaggerate its severity, and blame others for the impacts. This was done through fake science, social media manipulation, and the manipulation of the WHO.

Notes

1 Norman M. Naimark, *Stalin's Genocides* (Princeton, NJ: Princeton University Press, 2010), 99–120.
2 Robert C. Tucker, *Stalin in Power: The Revolution from Above, 1928–1941* (New York: W.W. Norton, 1992), 272.
3 Tucker, *Stalin in Power*, 445.
4 Tucker, *Stalin in Power*, 272.
5 Tucker, *Stalin in Power*, 272.
6 Tucker, *Stalin in Power*, 273.
7 Matt Lenoe, 'Did Stalin Kill Kirov and Does It Matter?' *The Journal of Modern History* 74, no. 2 (2002): 352–380.
8 Donald Rayfield, *Stalin and His Hangmen: An Authoritative Portrait of a Tyrant and Those Who Served Him* (London: Viking, 2004), 240.
9 Rayfield, *Stalin and His Hangmen*, 239–240.
10 Rayfield, *Stalin and His Hangmen*, 243.
11 Tucker, *Stalin in Power*, 318.
12 Tucker, *Stalin in Power*, 371.
13 Rayfield, *Stalin and His Hangmen*, 273.
14 Barry McLoughlin and Kevin McDermott, 'Rethinking Stalinist Terror,' in: Barry McLoughlin and Kevin McDermott (eds.), *Stalin's Terror: High Politics and Mass Repression in the Soviet Union* (New York: Palgrave Macmillan, 2003), 2.
15 Tucker, *Stalin in Power*, 433.
16 Rayfield, *Stalin and His Hangmen*, 313.
17 Rayfield, *Stalin and His Hangmen*, 314–315.
18 Rayfield, *Stalin and His Hangmen*, 315.
19 Barry McLoughlin, 'Mass Operations of the NKVD, 1937–8: A Survey,' in: Barry McLoughlin and Kevin McDermott (eds.), *Stalin's Terror: High Politics and Mass Repression in the Soviet Union* (New York: Palgrave Macmillan, 2003), 119.
20 Tucker, *Stalin in Power*, 458.
21 Tucker, *Stalin in Power*, 442.
22 James Harris, *The Great Fear: Stalin's Terror of the 1930s* (Oxford: Oxford University Press, 2016), 1.
23 Oleg Khlevniuk, 'Party and NKVD: Power Relationships in the Years of the Great Terror,' in: Barry McLoughlin and Kevin McDermott (eds.), *Stalin's Terror: High Politics and Mass Repression in the Soviet Union* (New York: Palgrave Macmillan, 2003), 22–23.
24 Khlevniuk, 'Party and NKVD,' 27–28.

25 Khlevniuk, 'Party and NKVD,' 30.
26 Rayfield, *Stalin and His Hangmen*, 329.
27 Rayfield, *Stalin and His Hangmen*, 328–329.
28 Tucker, *Stalin in Power*, 443–444.
29 Mark Safranski, '5GW: Into the Heart of Darkness,' in: Abbot (ed.), *The Handbook of 5GW*, 129.
30 Eric Schmidt and Jared Cohen, *The New Digital Age: Reshaping the Future of People, Nations and Business* (New York: Alfred Knopf, 2013), 121–128.
31 Ken Dilanian, 'U.S. Intelligence Official Acknowledges Missed Arab Spring Signs,' *Los Angeles Times*, July 19, 2012.
32 Emily A. Zerndt, *The House That Propaganda Built: Historicizing the Democracy Promotion Efforts and Measurement Tools of Freedom House*, PhD dissertation, submitted to Western Michigan University, August 2016, 1, 64.
33 Freedom House, 'Financial Statements Year Ended June 30, 2021 and Independent Auditors' Report,' Freedom House website, available at: https://freedomhouse.org/sites/default/files/2021-11/Freedom House Final Report.pdf, accessed May 24, 2023.
34 Zerndt, *The House That Propaganda Built*, 221.
35 Gene Sharp, *The Politics of Nonviolent Action* (Boston, MA: Porter Sargent, 1974).
36 David Ignatius, 'Innocence Abroad: The New World of Spyless Coups,' *Washington Post*, September 22, 1991.
37 William Blum, *Rogue State: A Guide to the World's Only Superpower* (Monroe, ME: Common Courage Press, 2000).
38 Philip Agee, 'Tracking Covert Action into the Future,' *Covert Action Information Bulletin* 42 (Fall 1992), available at: http://mediafilter.org/MFF/CovOps.html
39 Seth Jones, *A Covert Action: Reagan, the CIA, and the Cold War Struggle in Poland* (New York: W.W. Norton, 2018), 304.
40 Jeremy M. Sharp, 'The Middle East Partnership Initiative: An Overview,' Congressional Research Service, February 8, 2005.
41 Scott Lindlaw, 'Bush: Iraq War Will Build Mideast Peace,' *Cincinnati Inquirer*, February 27, 2003.
42 George W. Bush, 'Remarks by President George W. Bush at the 20th Anniversary of the National Endowment for Democracy, November 6, 2003, available at: www.ned.org/remarks-by-president-george-w-bush-at-the-20th-anniversary/
43 'The Greater Middle East Initiative: Implementing a Vision,' *Strategic Comments* 12, no. 2 (2004), 1.
44 Mark Landler, 'Secret Report Ordered by Obama Identified Potential Arab Uprisings,' *New York Times*, February 17, 2011.
45 Landler, 'Secret Report Ordered by Obama Identified Potential Arab Uprisings.'
46 Erin A. Snider and David M. Faris, 'The Arab Spring: US Democracy Promotion in Egypt,' *The Middle East Policy Council* 18, no. 3 (2011): 50.
47 Emad Mekay, 'Exclusive: US Bankrolled anti-Morsi Activists,' *Al Jazeera*, July 10, 2013, available at: www.aljazeera.com/indepth/features/2013/07/2013710113522489801.html
48 Mekay, 'Exclusive.'
49 Mekay, 'Exclusive.'
50 Jay Newton-Small, 'Hillary's Little Startup: How the U.S. Is Using technology to Aid Syria's Rebels,' *Time*, June 13, 2012, available at: http://world.time.com/2012/06/13/hillarys-little-startup-how-the-u-s-is-using-technology-to-aid-syrias-rebels/
51 Ron Nixon, 'U.S. Groups Helped Nurture Arab Uprisings,' *New York Times*, April 14, 2011.
52 Deborah Amos, 'For Some Arab Revolutionaries, a Serbian Tutor', *NPR*, December 13, 2011, available at: www.npr.org/2011/12/13/143648877/for-some-arab-revolutionaries-a-serbian-tutor
53 Judy Bachrach, 'Wikihistory: Did the Leaks Inspire the Arab Spring?' *World Affairs* 174, no. 2 (2011): 41.
54 Daniel Estulin, *Deconstructing WikiLeaks* (Walterville, OR: Trine Day, 2012), 14–15.

55 Merlyna Lim, 'Framing Bouazizi: "White Lies", Hybrid Network, and Collective/ Connective Action in the 2010–11 Tunisian Uprising,' *Journalism* 14, no. 7 (2013): 927–928.

56 Alexis C. Madrigal, 'The Inside Story How Facebook Responded to Tunisian Hacks,' *The Atlantic*, January 24, 2011, available at: www.theatlantic.com/technology/archive/ 2011/01/the-inside-story-of-how-facebook-responded-to-tunisian-hacks/70044/

57 Jose Antonio, 'Spring Awakening: How an Egyptian Revolution Began on Facebook,' *New York Times*, February 17, 2012, available at: www.nytimes.com/2012/02/19/books/ review/how-an-egyptian-revolution-began-on-facebook.html?pagewanted=all&_r=1&

58 Wikileaks, *Global Intelligence Files*, email from Anya Alfano to Fred Burton from February 14, 2011, Document Nr. 5376052.

59 Agence France-Presse, 'US Trains Activists to Evade Security Forces,' Agence France-Presse, April 8, 2011, available at: www.rawstory.com/rs/2011/04/08/us-trains-activists-to-evade-security-forces/

60 James Glanz and John Markoff, 'U.S. Underwrites Internet Detour around Censors,' *New York Times*, June 12, 2011, available at: www.nytimes.com/2011/06/12/ world/12internet.html?pagewanted=all&_r=0

61 Jared Cohen, 'Jared Cohen at the NExTWORK Conference 2011 Interviewed by Steven Levy,' June 22, 2011, available at: www.youtube.com/watch?v=s9NDox06MjI

62 Schmidt and Cohen, *The New Digital Age*.

63 David Dayen, 'The Android Administration: Google's Remarkably Close Relationship with the Obama White House,' *The Intercept*, April 22, 2016, available at: https:// theinercept.com/2016/04/22/googles-remarkably-close-relationship-with-the-obama-white-house-in-two-charts/

64 Tim Ross, Matthew Moore, and Steven Swinford, 'Egypt Protests: America's Secret Backing for Rebel Leaders behind Uprising,' *The Daily Telegraph*, January 28, 2011, available at: www.telegraph.co.uk/news/worldnews/africaandindianocean/egypt/8289686/ Egypt-protests-Americas-secret-backing-for-rebel-leaders-behind-uprising.html

65 Hillary Clinton, 'Hillary Clinton's Remarks on Egyptian Protests,' *Washington Post*, January 28, 2011, www.washingtonpost.com/wp-dyn/content/article/2011/01/28/AR20 11012803722.html

66 Helene Cooper and Robert F. Worth, 'In Arab Spring Obama Finds a Sharp Test,' *New York Times*, September 24, 2012, available at: www.nytimes.com/2012/09/25/us/politics/ arab-spring-proves-a-harsh-test-for-obamas-diplomatic-skill.html?pagewanted=all

67 Valery Gerasimov, 'The Value of Science Is in the Foresight: New Challenges Demand Rethinking the Forms and Methods of Carrying Out Combat Operations,' *Military Review* (January–February 2016): 24.

68 Anthony Cordesman, 'Russia and the "Color Revolution": A Russian View of a World Destabilized by the US and the West,' Center for Strategic & International Studies, May 28, 2014.

69 Darya Korsunskaya, 'Putin Says, Russia Must Prevent "Color Revolution,"' Reuters, November 20, 2014, available at: www.reuters.com/article/us-russia-putin-security-idUSKCN0J41J620141120

70 Patricia Bromley, Evan Schofer, and Wesley Longhofer, 'Conventions over World Culture: The Rise of Legal Restrictions on Foreign Funding to NGOs, 1994–2015,' *Social Forces* 99, no. 1 (2020): 281–304.

71 F. Robles, and K. Semple, 'U.S. and Cuba Baffled by "Health Attacks" on American Envoys in Havana,' *New York Times*, August 12, 2017, A6.

72 A. Erickson, 'All the Theories What's Happening to the Diplomats in Cuba: At Least 21 Americans Have Reported a Bizarre Rash of Symptoms,' *Washington Post Blogs*, September 30, 2017, available at: www.washingtonpost.com/news/worldviews/wp/ 2017/09/30/all-the-theories-about-whats-happening-to-the-diplomats-in-cuba/

73 Associated Press, 'Mystery of Sonic Weapons Attacks at US Embassy in Cuba Deepens,' *The Guardian*, September 14, 2017, available at: www.theguardian.com/world/2017/sep/14/mystery-of-sonic-weapon-attacks-at-us-embassy-in-cuba-deepens

74 Robles and Semple, 'U.S. and Cuba Baffled by "Health Attacks" on American Envoys in Havana.'

75 J. Ioffe, 'The Mystery of the Immaculate Concussion,' *GQ Magazine*, October 20, 2020, available at: www.gq.com/story/cia-investigation-and-russian-microwave-attacks

76 H. Gardiner, '25th Person at U.S. Embassy Is Mysteriously Sickened,' *New York Times*, June 21, 2018, available at: www.nytimes.com/2018/06/21/us/politics/us-diplomat-cuba-embassy-illness.html

77 S.L. Meyers and J. Perlez, 'U.S. Diplomats Evacuated in China as Medical Mystery Grows,' *New York Times*, June 6, 2018.

78 Jack Dutton, 'Havana Syndrome Symptoms Have Been Reported in these Countries,' *Newsweek*, September 21, 2021, available at: www.newsweek.com/havana-syndrome-china-austria-russia-1631102

79 Paul Kolbe, Marc Polymeropoulus, and John Sipher, 'Havana and the Global Hunt for U.S. Officers,' *The Cipher Brief*, October 24, 2021, available at: www.thecipherbrief.com/havana-syndrome

80 U.S. Congress, H.R.4914 passed on August 3, 2021, available at: www.congress.gov/117/bills/hr4914/BILLS-117hr4914ih.pdf, Section 2 (13).

81 Ken Dilanian, '"Havana Syndrome" Symptoms in Small Group Most Likely Caused by Directed Energy, Says U.S. Intel Panel of Experts,' *NBC News*, February 2, 2022, available at: www.nbcnews.com/politics/national-security/havana-syndrome-symptoms-small-group-likely-caused-directed-energy-say-rcna14584

82 Randel L. Swanson II, Stephen Hampton, Judith Green-McKenzie, Ramon Diaz-Arrastia, M. Sean Grady, Ragini Verma, Rosette Biester, Diana Duda, Ronald L. Wolf, and Douglas H. Smith, 'Neurological Manifestations among US Government Personnel Reporting Directional Audible and Sensory Phenomena in Havana,' *JAMA* 319, no. 11 (2018): 1125–1133.

83 R. Verma, Randel L. Swanson, D. Parker, A.A. Ould Ismail, R.T. Shinohara, J.A. Alapatt, and J. Doshi, 'Neuroimaging Findings in U.S. Government Personnel with Possible Exposure to Directional Phenomena in Havana, Cuba,' *JAMA* 322, no. 4 (2019): 336–347. doi:10.1001/jama.2019.9269

84 W. Broad, 'Microwave Weapons Are Prime Suspect in Ills of U.S. Embassy Workers,' *New York Times*, September 1, 2018.

85 Broad, 'Microwave Weapons Are Prime Suspect in Ills of U.S. Embassy Workers.'

86 National Academies of Sciences, 'New Report Assesses Illnesses among U.S. Government Personnel and Their Families at Overseas Embassies,' *Engineering Medicine News Release*, December 5, 2020, available at: www.nationalacademies.org/news/2020/12/new-report-assesses-illnesses-among-us-government-personnel-and-their-families-at-overseas-embassies

87 David A. Relman and Julie A. Pavlin (eds.), *An Assessment of Illness in U.S. Government Employees and Their Families in Overseas Embassies* (Washington, DC: National Academies Press, 2020), 28.

88 Relman and Pavlin (eds.), *An Assessment of Illness in U.S. Government Employees and Their Families in Overseas Embassies*, 18.

89 Relman and Pavlin (eds.), *An Assessment of Illness in U.S. Government Employees and Their Families in Overseas Embassies*, 18.

90 Reuters, '"Havana syndrome" U.S. Diplomats Get Benefits in Spending Bill,' Reuters, December 16, 2019, available at: www.reuters.com/article/us-usa-budget-congress-diplomats/havana-syndrome-u-s-diplomats-get-benefits-in-spending-bill-idUSKBN1YK24Z

91 Kylie Atwood, Katie Bo Lillis, and Jennifer Hansler, 'Biden Administration to Compensate Some "Havana Syndrome" Victims up to $187,000,' CNN, June 24, 2022, available at: www.cnn.com/2022/06/23/politics/havana-syndrome-victims-compensation/index.html
92 U.S. Congress, H.R.4914.
93 U.S. Congress, H.R.4914, Section 2 (1).
94 U.S. Congress, H.R.4914, Section 2 (13).
95 Office of the Director of National Intelligence, 'Anomalous Health Incidents: Analysis of Potential Causal Mechanisms,' IC Experts Panel, September 2022, 14.
96 Office of the Director of National Intelligence, 'Anomalous Health Incidents,' 18.
97 Office of the Director of National Intelligence, 'Updated Assessment of Anomalous Health Incidents,' National Intelligence Council, March 1, 2023.
98 Office of the Director of National Intelligence, 'Updated Assessment of Anomalous Health Incidents.'
99 Tom Rogan, 'US Intelligence Community Has Proven It Can't Investigate Havana Syndrome, It Should Let UK and Australia Try,' *Washington Examiner*, March 1, 2023, availableat:www.washingtonexaminer.com/opinion/the-us-intelligence-community-has-proven-it-cant-investigate-havana-syndrome-it-should-let-the-uk-and-australia-try
100 Rogan, 'US Intelligence Community Has Proven It Can't Investigate Havana Syndrome, It Should Let UK and Australia Try.'
101 David Ignatius, 'Dealing with "Havana Syndrome" Is a Policymaker's Nightmare,' *Washington Post*, October 28. 2021.
102 Julian Borger, 'Havana Syndrome Has "Dramatically Hurt" Morale, US Diplomats Say,' *The Guardian*, February 10, 2022, available at: www.theguardian.com/us-news/2022/feb/10/havana-syndrome-cuba-us-diplomats-afsa
103 Alina Chan and Matt Ridley, *Viral: The Search for the Origin of COVID-19* (New York: HarperCollins, 2021), 50.
104 Pete Hoekstra and William Boykin, *The CCP Is at War with America: A Team B Report on the Covid 19 Biological Warfare Attack* (Washington, DC: Center for Security Policy, 2022), 68.
105 World Health Organization, 'WHO Timeline—COVID-19,' Statement, April 27, 2020, available at: www.who.int/news/item/27-04-2020-who-timeline---covid-19
106 Emma Parker, 'Coronavirus Blamed on Bat Soup as Pics Emerge of Residents Eating Unusual Cuisine,' *Daily Star*, January 22, 2020.
107 Former CDC Director Redfield stated that he even received death threats from scientists for publicly stating his belief that COVID-19 emerged from a lab. Edmund DeMarche, 'Ex-CDC Director Says He Received Death Threats after Mentioning Lab-Leak Theory,' *Fox News*, June 3, 2021, available at: www.foxnews.com/health/ex-cdc-director-redfield-says-he-received-death-threats-after-mentioning-lab-leak-theory
108 M. Anthony Mills, 'Manufacturing Consensus,' *The New Atlantis* 66 (Fall 2021): 30.
109 Jeffrey Sachs, 'Why the Chair of The Lancet's COVID-19 Commission Thinks the US Government Is Preventing a Real Investigation into the Pandemic,' *Current Affairs*, August 2, 2022, available at: www.currentaffairs.org/2022/08/why-the-chair-of-the-lancets-covid-19-commission-thinks-the-us-government-is-preventing-a-real-investigation-into-the-pandemic
110 Chan and Ridley, *Viral*, 10–20, 133–141.
111 U.S. Department of State, 'Fact Sheet: Activity at the Wuhan Institute of Virology,' Office of the Spokesperson, January 15, 2021, available at: https://2017-2021.state.gov/fact-sheet-activity-at-the-wuhan-institute-of-virology/index.html
112 Bob Kadlec and Bob Foster, 'Muddy Waters: The Origins of COVID-19 Report,' U.S. Senate, April 17, 2023, 17.
113 Reuters, 'Former CDC Chief Redfield Says He Thinks COVID-19 Originated in a Chinese Lab,' Reuters, March 26, 2021, available at: www.reuters.com/business/healthcare-pharmaceuticals/former-cdc-chief-redfield-says-he-thinks-covid-19-orig

inated-chinese-lab-2021–03–26/; Ethan Ennals, 'Trump Aide Claims Covid "Came Out of the Box Ready to Infect"—Claiming Virus Was Being Worked on by Scientists in a Chinese Lab,' *Daily Mail*, July 16, 2022, available at: www.dailymail.co.uk/news/arti cle-11021329/Trump-aide-claims-Covid-came-box-ready-infect.html; Animuta Kaur and Dan Diamond, 'FBI Director Says COVID-19 "Most Likely" Emerged from a Lab,' *Washington Post*, February 28, 2023.

114 Li-Meng Yan, Shu Kang, and Shanchang Hu, 'SARS-CoV-2 Is an Unrestricted Bio- weapon: A Truth Revealed through Uncovering a Large-Scale, Organized Scientific Fraud,' *Zenodo* (October 2020), doi:10.5281/zenodo.4073130

115 Yan, Kang, and Hu, 'SARS-CoV-2 Is an Unrestricted Bioweapon,' 27.

116 U.S. Department of State, 'Fact Sheet.'

117 Michael McCaul, 'The Origins of COVID-19: An Investigation of the Wuhan Institute of Virology,' House Foreign Affairs Committee Report Minority Staff, August 2021, 17.

118 Brandon J. Weichert, *Biohacked: China's Race to Control Life* (New York: Encounter Books, 2023), 37–38.

119 Hoekstra and Boykin, *The CCP Is at War With America*, 102–103.

120 Chan and Ridley, *Viral*, 245.

121 Chan and Ridley, *Viral*, 258.

122 McCaul, 'The Origins of COVID-19,' 46.

123 According to Brandon Weichert, WHO Director Tedros Adhanom Ghebreyesus pri- vately voiced the suspicion that SARS-CoV-2 originated from the WIV while publicly refusing to acknowledge that a lab leak is the likely origin. Weichert, *Biohacked*, 37.

124 Maria Bartiromo, 'Tim Cotton Explains Why He Thinks China Allowed Flights out of Wuhan during Outbreak,' *The Daily Wire*, April 26, 2020, available at: www.dailywire. com/news/tom-cotton-explains-why-he-thinks-china-allowed-flights-out-of-wuhan- during-outbreak

125 JonRoss Wendler, 'Misleading a Pandemic: The Viral Effects of Chinese Propaganda and the Coronavirus,' *Joint Forces Quarterly* 104, no. 1 (2022): 37.

126 Steven Lee Myers, 'China Spins Tale That the U.S. Army Started the Coronavirus Epidemic,' *New York Times*, March 13, 2020.

127 Ben Nimmo, Camille Francois, C. Shawn Eib, and Léa Ronzaud, 'Spamouflage Dragon Goes to America: Pro-Chinese Inauthentic Network Debuts English-Language Videos,' *Graphika* (August 2020), available at: https://public-assets.graphika.com/ reports/graphika_report_spamouflage_dragon_goes_to_america.pdf, 30–31.

128 Edward Wong, Matthew Rosenberg, and Julian Barnes, 'Chinese Agents Helped Spread Messages That Sowed Panic in U.S., Officials Say,' *New York Times*, April 23, 2020.

129 Michael P. Senger, 'China's Global Lockdown Propaganda Campaign,' *Tablet Maga- zine*, September 15, 2020, available at: www.tabletmag.com/sections/news/articles/ china-covid-lockdown-propaganda

130 Senger, 'China's Global Lockdown Propaganda Campaign.'

131 Nimmo et al., 'Spamouflage Dragon Goes to America,' 30–31.

132 Weichert, *Biohacked*, 56–60.

133 Yi Rao, 'My Relatives in Wuhan Survived. My Uncle in New York Did Not,' *New York Times*, July 23, 2020.

134 Weichert, *Biohacked*, 91.

135 Worldometer, 'Coronavirus Statistics,' website, available at: www.worldometers.info/ coronavirus/, accessed on May 20, 2023.

136 Rebekah Levine Coley, Naoka Carey, Christopher F. Baum, and Summer Sherburne Hawkins, 'COVID-19-Related Stressors and Mental Health Disorders among US Adults,' *Public Health Reports* 137, no, 6 (2022): 1217–1226.

137 Government Accountability Office, 'COVID-19 Relief: Funding and Spending as of Jan. 31, 2023,' GAO-23–106647, February 2023, available at: www.gao.gov/assets/ gao-23-106647.pdf

138 Jakub Hlávka and Adam Rose, 'Opinion: "Unprecedented Costs by Most Measures": Calculating the Astonishing Economic Costs of Covid,' *Los Angeles Times*, May 15, 2023, available at: www.latimes.com/opinion/story/2023-05-15/covid-pandemic-us-economic-costs-14-trillion
139 U.S. Department of the Treasury, 'Executive Summary to the Fiscal Year 2022 Financial Report of U.S. Government: An Unsustainable Fiscal Path,' Bureau of the Fiscal Service, April 6, 2023, available at: www.gao.gov/assets/gao-23-106647.pdf
140 Mark Cudmore, 'US Annualized Debt Now Exceeds $800 Billion,' Zerohedge, April 17, 2023, available at: www.zerohedge.com/markets/us-annualized-debt-costs-exceed-800-billion
141 Hoekstra and Boykin, *The CCP Is at War With America*, 7.
142 Lewis Pennock, 'Supreme Court Justice Gorsuch Issues Excoriating Review of COVID Lockdown Policies Including Business Closures and Vaccine Mandates and Calls Them "Among the Greatest Intrusions on Civil Liberties in the History of the Nation,"' *Daily Mail*, May 20, 2023, available at: www.dailymail.co.uk/news/article-12106351/Supreme-Court-justice-tears-COVID-lockdown-vaccine-policies.html
143 Richard Fleming, *Is COVID-19 a Bioweapon? A Scientific and Forensic Investigation* (New York: Skyhorse, 2021), 101.
144 Bill Gates suggested that the next pandemic would be caused by bioterrorism and that it would be more deadly than COVID-19. Joseph Guzman, 'Bill Gates, Who Predicted the Pandemic, Names the Next Two Monsters That Could Shake Our World,' *The Hill*, February 11, 2021, available at: https://thehill.com/changing-america/well-being/538426-bill-gates-who-predicted-the-pandemic-names-the-next-two-monster/

References

Agee, Philip, 'Tracking Covert Action into the Future,' *Covert Action Information Bulletin* 42 (Fall 1992), available at: http://mediafilter.org/MFF/CovOps.html
Agence France-Presse, 'US Trains Activists to Evade Security Forces,' Agence France-Presse, April 8, 2011, available at: www.rawstory.com/rs/2011/04/08/us-trains-activists-to-evade-security-forces/
Amos, Deborah, 'For Some Arab Revolutionaries, a Serbian Tutor', *NPR*, December 13, 2011, available at: www.npr.org/2011/12/13/143648877/for-some-arab-revolutionaries-a-serbian-tutor
Antonio, Jose, 'Spring Awakening: How an Egyptian Revolution Began on Facebook,' *New York Times*, February 17, 2012, available at: www.nytimes.com/2012/02/19/books/review/how-an-egyptian-revolution-began-on-facebook.html?pagewanted=all&_r=1&
Associated Press, 'Mystery of Sonic Weapons Attacks at US Embassy in Cuba Deepens,' *The Guardian*, September 14, 2017, available at: www.theguardian.com/world/2017/sep/14/mystery-of-sonic-weapon-attacks-at-us-embassy-in-cuba-deepens
Atwood, Kylie, Lillis, Katie Bo, and Hansler, Jennifer, 'Biden Administration to Compensate Some "Havana Syndrome" Victims up to $187,000,' CNN, June 24, 2022, available at: www.cnn.com/2022/06/23/politics/havana-syndrome-victims-compensation/index.html
Bachrach, Judy, 'Wikihistory: Did the Leaks Inspire the Arab Spring?' *World Affairs* 174, no. 2 (2011): 35–44.
Bartiromo, Maria, 'Tim Cotton Explains Why He Thinks China Allowed Flights out of Wuhan during Outbreak,' *The Daily Wire*, April 26, 2020, available at: www.dailywire.com/news/tom-cotton-explains-why-he-thinks-china-allowed-flights-out-of-wuhan-during-outbreak
Blum, William, *Rogue State: A Guide to the World's Only Superpower* (Monroe, ME: Common Courage Press, 2000).

Borger, Julian, 'Havana Syndrome Has "Dramatically Hurt" Morale, US Diplomats Say,' *The Guardian*, February 10, 2022, available at: www.theguardian.com/us-news/2022/feb/10/havana-syndrome-cuba-us-diplomats-afsa

Broad, W., 'Microwave Weapons Are Prime Suspect in Ills of U.S. Embassy Workers,' *New York Times*, September 1, 2018.

Bromley, Patricia, Schofer, Evan, and Longhofer, Wesley, 'Conventions over World Culture: The Rise of Legal Restrictions on Foreign Funding to NGOs, 1994–2015,' *Social Forces* 99, no. 1 (2020): 281–304.

Bush, George W., 'Remarks by President George W. Bush at the 20th Anniversary of the National Endowment for Democracy,' November 6, 2003, available at: www.ned.org/remarks-by-president-george-w-bush-at-the-20th-anniversary/

Chan, Alina and Ridley, Matt, *Viral: The Search for the Origin of COVID-19* (New York: HarperCollins, 2021).

Clinton, Hillary, 'Hillary Clinton's Remarks on Egyptian Protests,' *Washington Post*, January 28, 2011, www.washingtonpost.com/wp-dyn/content/article/2011/01/28/AR2011012803722.html

Cohen, Jared, 'Jared Cohen at the NExTWORK Conference 2011 Interviewed by Steven Levy,' June 22, 2011, available at: www.youtube.com/watch?v=s9NDox06MjI

Cooper, Helene and Worth, Robert F., 'In Arab Spring Obama Finds a Sharp Test,' *New York Times*, September 24, 2012, available at: www.nytimes.com/2012/09/25/us/politics/arab-spring-proves-a-harsh-test-for-obamas-diplomatic-skill.html?pagewanted=all

Cordesman, Anthony, 'Russia and the "Color Revolution": A Russian View of a World Destabilized by the US and the West,' Center for Strategic & International Studies, May 28, 2014.

Cudmore, Mark, 'US Annualized Debt Now Exceeds $800 Billion,' Zerohedge, April 17, 2023, available at: www.zerohedge.com/markets/us-annualized-debt-costs-exceed-800-billion

Dayen, David, 'The Android Administration: Google's Remarkably Close Relationship with the Obama White House,' *The Intercept*, April 22, 2016, available at: https://theintercept.com/2016/04/22/googles-remarkably-close-relationship-with-the-obama-white-house-in-two-charts/

DeMarche, Edmund, 'Ex-CDC Director Says He Received Death Threats after Mentioning Lab-Leak Theory,' *Fox News*, June 3, 2021, available at: www.foxnews.com/health/ex-cdc-director-redfield-says-he-received-death-threats-after-mentioning-lab-leak-theory

Dilanian, Ken, 'U.S. Intelligence Official Acknowledges Missed Arab Spring Signs,' *Los Angeles Times*, July 19, 2012.

———, '"Havana Syndrome" Symptoms in Small Group Most Likely Caused by Directed Energy, Says U.S. Intel Panel of Experts,' *NBC News*, February 2, 2022, available at: www.nbcnews.com/politics/national-security/havana-syndrome-symptoms-small-group-likely-caused-directed-energy-say-rcna14584

Dutton, Jack, 'Havana Syndrome Symptoms Have Been Reported in these Countries,' *Newsweek*, September 21, 2021, available atwww.newsweek.com/havana-syndrome-china-austria-russia-1631102

Ennals, Ethan, 'Trump Aide Claims Covid "Came Out of the Box Ready to Infect"—Claiming Virus Was Being Worked on by Scientists in a Chinese Lab,' *Daily Mail*, July 16, 2022, available at: www.dailymail.co.uk/news/article-11021329/Trump-aide-claims-Covid-came-box-ready-infect.html

Erickson, A., 'All the Theories about What's Happening to the Diplomats in Cuba: At Least 21 Americans Have Reported a Bizarre Rash of Symptoms,' *Washington Post*

Blogs, September 30, 2017, available at: www.washingtonpost.com/news/worldviews/wp/2017/09/30/all-the-theories-about-whats-happening-to-the-diplomats-in-cuba/

Estulin, Daniel, *Deconstructing WikiLeaks* (Walterville, OR: Trine Day, 2012).

Fleming, Richard, *Is COVID-19 a Bioweapon? A Scientific and Forensic Investigation* (New York: Skyhorse, 2021).

Freedom House, 'Financial Statements Year Ended June 30, 2021 and Independent Auditors' Report,' Freedom House website, available at: https://freedomhouse.org/sites/default/files/2021-11/Freedom House Final Report.pdf, accessed May 24, 2023.

Gardiner, H., '25th Person at U.S. Embassy Is Mysteriously Sickened,' *New York Times*, June 21, 2018, available at: www.nytimes.com/2018/06/21/us/politics/us-diplomat-cuba-embassy-illness.html

Gerasimov, Valery, 'The Value of Science Is in the Foresight: New Challenges Demand Rethinking the Forms and Methods of Carrying out Combat Operations,' *Military Review* (January–February 2016): 23–29.

Glanz, James and Markoff, John, 'U.S. Underwrites Internet Detour around Censors,' *New York Times*, June 12, 2011, available at: www.nytimes.com/2011/06/12/world/12internet.html?pagewanted=all&_r=0

Government Accountability Office, 'COVID-19 Relief: Funding and Spending as of Jan. 31, 2023,' GAO-23–106647, February 2023, available at: www.gao.gov/assets/gao-23-106647.pdf

Guzman, Joseph, 'Bill Gates, Who Predicted the Pandemic, Names the Next Two Monsters That Could Shake Our World,' *The Hill*, February 11, 2021, available at: https://thehill.com/changing-america/well-being/538426-bill-gates-who-predicted-the-pandemic-names-the-next-two-monster/

Harris, James, *The Great Fear: Stalin's Terror of the 1930s* (Oxford: Oxford University Press, 2016).

Hlávka, Jakub and Rose, Adam, 'Opinion: "Unprecedented Costs by Most Measures": Calculating the Astonishing Economic Costs of Covid,' *Los Angeles Times*, May 15, 2023, available at: www.latimes.com/opinion/story/2023-05-15/covid-pandemic-us-economic-costs-14-trillion

Hoekstra, Pete and Boykin, William, *The CCP Is at War with America: A Team B Report on the Covid 19 Biological Warfare Attack* (Washington, DC: Center for Security Policy, 2022).

Ignatius, David, 'Innocence Abroad: The New World of Spyless Coups,' *Washington Post*, September 22, 1991.

———, 'Dealing with "Havana Syndrome" Is a Policymaker's Nightmare,' *Washington Post*, October 28. 2021.

Ioffe, J., 'The Mystery of the Immaculate Concussion,' *GQ Magazine*, October 20, 2020, available at: www.gq.com/story/cia-investigation-and-russian-microwave-attacks

Jones, Seth, *A Covert Action: Reagan, the CIA, and the Cold War Struggle in Poland* (New York: W.W. Norton, 2018).

Kadlec, Bob and Foster, Bob, 'Muddy Waters: The Origins of COVID-19 Report,' U.S. Senate, April 17, 2023.

Kaur, Animuta and Diamond, Dan, 'FBI Director Says COVID-19 "Most Likely" Emerged from a Lab,' *Washington Post*, February 28, 2023.

Khlevniuk, Oleg, 'Party and NKVD: Power Relationships in the Years of the Great Terror,' in: Barry McLoughlin and Kevin McDermott (eds.), *Stalin's Terror: High Politics and Mass Repression in the Soviet Union* (New York: Palgrave Macmillan, 2003), 21–33.

Kolbe, Paul, Polymeropoulus, Marc, and Sipher, John, 'Havana and the Global Hunt for U.S. Officers,' *The Cipher Brief*, October 24, 2021, available at: www.thecipherbrief.com/havana-syndrome

Korsunskaya, Darya, 'Putin Says, Russia Must Prevent "Color Revolution,"' Reuters, November 20, 2014, available at: www.reuters.com/article/us-russia-putin-security-id USKCN0J41J620141120

Landler, Mark, 'Secret Report Ordered by Obama Identified Potential Arab Uprisings,' *New York Times*, February 17, 2011.

Lenoe, Matt, 'Did Stalin Kill Kirov and Does It Matter?' *The Journal of Modern History* 74, no. 2 (2002): 352–380.

Levine Coley, Rebekah, Carey, Naoka, Baum, Christopher F., and Sherburne Hawkins, Summer, 'COVID-19-Related Stressors and Mental Health Disorders among US Adults,' *Public Health Reports* 137, no, 6 (2022): 1217–1226.

Lim, Merlyna, 'Framing Bouazizi: "White Lies", Hybrid Network, and Collective/Connective Action in the 2010–11 Tunisian Uprising,' *Journalism* 14, no. 7 (2013): 921–941.

Lindlaw, Scott, 'Bush: Iraq War Will Build Mideast Peace,' *Cincinnati Inquirer*, February 27, 2003.

Madrigal, Alexis C., 'The Inside Story How Facebook Responded to Tunisian Hacks,' *The Atlantic*, January 24, 2011, available at: www.theatlantic.com/technology/archive/2011/01/the-inside-story-of-how-facebook-responded-to-tunisian-hacks/70044/

McCaul, Michael, 'The Origins of COVID-19: An Investigation of the Wuhan Institute of Virology,' House Foreign Affairs Committee Report Minority Staff, August 2021.

McLoughlin, Barry, 'Mass Operations of the NKVD, 1937–8: A Survey,' in: Barry McLoughlin and Kevin McDermott (eds.), *Stalin's Terror: High Politics and Mass Repression in the Soviet Union* (New York: Palgrave Macmillan, 2003), 118–152.

——— and McDermott, Kevin, 'Rethinking Stalinist Terror,' in: Barry McLoughlin and Kevin McDermott (eds.), *Stalin's Terror: High Politics and Mass Repression in the Soviet Union* (New York: Palgrave Macmillan, 2003), 1–18.

Mekay, Emad, 'Exclusive: US Bankrolled anti-Morsi Activists,' Al Jazeera, July 10, 2013, available at: www.aljazeera.com/indepth/features/2013/07/2013710113522489801.html

Meyers, S.L. and Perlez, J., 'U.S. Diplomats Evacuated in China as Medical Mystery Grows,' *New York Times*, June 6, 2018.

Mills, M. Anthony, 'Manufacturing Consensus,' *The New Atlantis* 66 (Fall 2021): 30–45.

Myers, Steven Lee, 'China Spins Tale That the U.S. Army Started the Coronavirus Epidemic,' *New York Times*, March 13, 2020.

Naimark, Norman M., *Stalin's Genocides* (Princeton, NJ: Princeton University Press, 2010).

National Academies of Sciences, 'New Report Assesses Illnesses among U.S. Government Personnel and Their Families at Overseas Embassies,' *Engineering Medicine News Release*, December 5, 2020, available at: www.nationalacademies.org/news/2020/12/new-report-assesses-illnesses-among-us-government-personnel-and-their-families-at-overseas-embassies

Newton-Small, Jay, 'Hillary's Little Startup: How the U.S. Is Using Technology to Aid Syria's Rebels,' *Time*, June 13, 2012, available at: http://world.time.com/2012/06/13/hillarys-little-startup-how-the-u-s-is-using-technology-to-aid-syrias-rebels/

Nimmo, Ben, Francois, Camille, Eib, C. Shawn, and Ronzaud, Léa, 'Spamouflage Dragon Goes to America: Pro-Chinese Inauthentic Network Debuts English-Language Videos,' *Graphika* (August 2020), available at: https://public-assets.graphika.com/reports/graphika_report_spamouflage_dragon_goes_to_america.pd

Nixon, Ron, 'U.S. Groups Helped Nurture Arab Uprisings,' *New York Times*, April 14, 2011.

Office of the Director of National Intelligence, 'Anomalous Health Incidents: Analysis of Potential Causal Mechanisms,' IC Experts Panel, September 2022.

———, 'Updated Assessment of Anomalous Health Incidents,' National Intelligence Council, March 1, 2023.

Parker, Emma, 'Coronavirus Blamed on Bat Soup as Pics Emerge of Residents Eating Unusual Cuisine,' *Daily Star*, January 22, 2020.

Pennock, Lewis, 'Supreme Court Justice Gorsuch Issues Excoriating Review of COVID Lockdown Policies Including Business Closures and Vaccine Mandates and Calls Them "Among the Greatest Intrusions on Civil Liberties in the History of the Nation,"' *Daily Mail*, May 20, 2023, available at: www.dailymail.co.uk/news/article-12106351/Supreme-Court-justice-tears-COVID-lockdown-vaccine-policies.html

Rao, Yi, 'My Relatives in Wuhan Survived. My Uncle in New York Did Not,' *New York Times*, July 23, 2020.

Rayfield, Donald, *Stalin and His Hangmen: An Authoritative Portrait of a Tyrant and Those Who Served Him* (London: Viking, 2004).

Relman, David A. and Pavlin, Julie A. (eds.), *An Assessment of Illness in U.S. Government Employees and Their Families in Overseas Embassies* (Washington, DC: National Academies Press, 2020).

Reuters, '"Havana syndrome" U.S. Diplomats Get Benefits in Spending Bill,' Reuters, December 16, 2019, available at: www.reuters.com/article/us-usa-budget-congress-diplomats/havana-syndrome-u-s-diplomats-get-benefits-in-spending-bill-idUSKBN1YK24Z

———, 'Former CDC Chief Redfield Says He Thinks COVID-19 Originated in a Chinese Lab,' Reuters, March 26, 2021, available at: www.reuters.com/business/healthcare-pharmaceuticals/former-cdc-chief-redfield-says-he-thinks-covid-19-originated-chinese-lab-2021-03-26/

Robles, F. and Semple, K., 'U.S. and Cuba Baffled by "Health Attacks" on American Envoys in Havana,' *New York Times*, August 12, 2017, A6.

Rogan, Tom, 'US Intelligence Community Has Proven It Can't Investigate Havana Syndrome, It Should Let UK and Australia Try,' *Washington Examiner*, March 1, 2023, available at: www.washingtonexaminer.com/opinion/the-us-intelligence-community-has-proven-it-cant-investigate-havana-syndrome-it-should-let-the-uk-and-australia-try

Ross, Tim, Moore, Matthew, and Swinford, Steven, 'Egypt Protests: America's Secret Backing for Rebel Leaders behind Uprising,' *The Daily Telegraph*, January 28, 2011, available at: www.telegraph.co.uk/news/worldnews/africaandindianocean/egypt/8289686/Egypt-protests-Americas-secret-backing-for-rebel-leaders-behind-uprising.html

Sachs, Jeffrey, 'Why the Chair of The Lancet's COVID-19 Commission Thinks the US Government Is Preventing a Real Investigation into the Pandemic,' *Current Affairs*, August 2, 2022, available at: www.currentaffairs.org/2022/08/why-the-chair-of-the-lancets-covid-19-commission-thinks-the-us-government-is-preventing-a-real-investigation-into-the-pandemic

Safranski, Mark, '5GW: Into the Heart of Darkness,' in: Daniel H. Abbot (ed.), *The Handbook of 5GW* (Ann Arbor, MI: Nimble Books, 2010), 125–135.

Schmidt, Eric and Cohen, Jared, *The New Digital Age: Reshaping the Future of People, Nations and Business* (New York: Alfred Knopf, 2013).

Senger, Michael P., 'China's Global Lockdown Propaganda Campaign,' *Tablet Magazine*, September 15, 2020, available at: www.tabletmag.com/sections/news/articles/china-covid-lockdown-propaganda

Sharp, Gene, *The Politics of Nonviolent Action* (Boston, MA: Porter Sargent, 1974).

Sharp, Jeremy M., 'The Middle East Partnership Initiative: An Overview,' Congressional Research Service, February 8, 2005.

Snider, Erin A. and Faris, David M., 'The Arab Spring: US Democracy Promotion in Egypt,' *The Middle East Policy Council* 18, no. 3 (2011): 49–62.

Swanson II, Randel L., Hampton, Stephen, Green-McKenzie, Judith, Diaz-Arrastia, Ramon, Grady, M. Sean, Verma, Ragini, Biester, Rosette, Duda, Diana, Wolf, Ronald L., Smith, and Douglas H., 'Neurological Manifestations among US Government Personnel Reporting Directional Audible and Sensory Phenomena in Havana,' *JAMA* 319, no. 11 (2018): 1125–1133.

'The Greater Middle East Initiative: Implementing a Vision,' *Strategic Comments* 12, no. 2 (2004), 1.

Tucker, Robert C., *Stalin in Power: The Revolution from Above, 1928–1941* (New York: W.W. Norton, 1992).

U.S. Congress, H.R.4914, passed on August 3, 2021, available at: www.congress.gov/117/bills/hr4914/BILLS-117hr4914ih.pdf, Section 2 (13).

U.S. Department of State, 'Fact Sheet: Activity at the Wuhan Institute of Virology,' Office of the Spokesperson, January 15, 2021, available at: https://2017-2021.state.gov/fact-sheet-activity-at-the-wuhan-institute-of-virology/index.html

U.S. Department of the Treasury, 'Executive Summary to the Fiscal Year 2022 Financial Report of U.S. Government: An Unsustainable Fiscal Path,' Bureau of the Fiscal Service, April 6, 2023, available at: www.gao.gov/assets/gao-23-106647.pdf

Verma, R., Swanson, Randel L., Parker, D., Ismail, A.A. Ould, Shinohara, R.T., Alapatt, J.A., and Doshi, J., 'Neuroimaging Findings in U.S. Government Personnel with Possible Exposure to Directional Phenomena in Havana, Cuba,' *JAMA* 322, no. 4 (2019): 336–347. doi:10.1001/jama.2019.9269

Weichert, Brandon J., *Biohacked: China's Race to Control Life* (New York: Encounter Books, 2023).

Wendler, JonRoss, 'Misleading a Pandemic: The Viral Effects of Chinese Propaganda and the Coronavirus,' *Joint Forces Quarterly* 104, no. 1 (2022): 32–39.

Wikileaks, *Global Intelligence Files*, email from Anya Alfano to Fred Burton from February 14, 2011, Document Nr. 5376052.

Wong, Edward, Rosenberg, Matthew, and Barnes, Julian, 'Chinese Agents Helped Spread Messages That Sowed Panic in U.S., Officials Say,' *New York Times*, April 23, 2020.

World Health Organization, 'WHO Timeline—COVID-19,' Statement, April 27, 2020, available at: www.who.int/news/item/27-04-2020-who-timeline---covid-19

Worldometer, 'Coronavirus Statistics,' website, available at: www.worldometers.info/coronavirus/, accessed on May 20, 2023.

Yan, Li-Meng, Kang, Shu, and Hu, Shanchang, 'SARS-CoV-2 Is an Unrestricted Bioweapon: A Truth Revealed through Uncovering a Large-Scale, Organized Scientific Fraud,' *Zenodo* (October 2020), doi:10.5281/zenodo.4073130

Zerndt, Emily A., *The House That Propaganda Built: Historicizing the Democracy Promotion Efforts and Measurement Tools of Freedom House*, PhD dissertation, submitted to Western Michigan University, August 2016.

5 Nano-Info-Bio-Cogno Technologies

So far, it has been established that 5GW is a mode of warfare at the societal level that takes place within the human domain and that targets the mind (perceptions and decision-making). Unlike the traditional view of war, defined as the instrumental use of violence aimed at coercing an enemy, 5GW theory posits that coercion may occur with a minimal amount of violence or with the help of violence that is covert or hidden from the victim or that is disguised as accidental or benign. This means that, like all war, 5GW is still violent, if not in terms of the means used, certainly in terms of the outcome that 5GW opponents seek to achieve. This can be political destabilization, even to the point of civil war, the complete political subjugation of a population, even to the point where any open resistance becomes impossible, or the covert, targeted elimination of a particular portion of a population (genocide or democide). It has been suggested that, although 5GW requires societal complexity and technology that did not previously exist in human history, there have been historical antecedents such as Stalin's Great Terror and Mao's Cultural Revolution. This chapter will discuss emerging technologies in the areas of nanotechnology, biotechnology, information technology, and cognitive or neurotechnology (NBIC) in relation to 5GW. NBIC technologies transform not only what humans can do but also, more importantly, human beings themselves. Transhumanists insist that humanity must be technologically upgraded in the pursuit of a paradisical post-human future. From a security and military perspective, the questions are: Who determines what would be the right direction for human evolution, and whose interests does it ultimately serve to modify humans in a specific manner? In short, transhumanism opens up the possibility of redesigning a society by way of redesigning human beings, which may lead to the technological enslavement of a designated underclass and/or the targeted, covert elimination of undesirables.

Transhumanism

Transhumanism is an old idea that is grounded in Darwin's theory of evolution (a proclaimed tendency of nature for the emergence of ever-more complex organisms) and that posits human evolution can continue by way of humans merging

DOI: 10.4324/9781003396963-6

with technology.[1] One of the early proponents of transhumanism was the biologist Julian Huxley, who claimed

> that even the most fortunate people are living far below capacity, and that most human beings develop not more than a small fraction of their potential mental and spiritual efficiency . . . the human species will be on the threshold of a new kind of existence, as different from ours as ours is from that of Peking man. It will at last be consciously fulfilling its real destiny.[2]

Transhumanism means to go beyond the natural limitations of man and potentially move towards a post-biological or post-human future of a world no longer inhabited by human beings but rather by superintelligent robots.[3]

The vision of transhumanists is now closer to reality than it has ever been owing to the convergence of several technology disciplines—*biotechnology*, or the manipulation of biological organisms; *information and computer technology*, or the ability to store, process, and analyze vast amounts of data; *nanotechnology*, or the manipulation of atoms and molecules; and *neurotechnology*, or the ability to understand brain functions and manipulate brain activity. Advances in each of these technology areas have helped to rapidly advance the other technology areas mentioned. This has made it possible for scientists and engineers to develop a much more comprehensive understanding of the mechanisms of life, cognition (AI), matter, and the human mind. As will be shown below, the technological possibilities are both extremely astonishing and terrifying if abused.

Biotechnology

The discovery of genetics, the sequencing of the human genome, and the development of cheap gene editing have created the possibility of human genetic engineering that not only fixes diseases and health conditions that have a genetic origin but also opens up the possibility of developing human physical and mental capabilities beyond natural limitations. Nick Bostrom suggested that transhumanists 'recognize that the promise of genetic enhancements is anything but insignificant. Being free from severe genetic diseases would be good, as would having a mind that can learn more quickly, or having a more robust immune system.'[4] For Bostrom, '[e]very day that the introduction of effective human genetic enhancement is delayed is a day of lost individual and cultural potential, and a day of torment for many unfortunate sufferers of diseases that could have been prevented.'[5] Ultimately, transhumanists want to conquer death, or at least greatly expand human life spans.[6]

However, Bostrom and other transhumanists are not content with the therapeutic application of genetic enhancement and suggest using genetic selection and stem cell-derived gametes to speed up evolution and produce superintelligent humans. Bostrom claimed that 'the average level of intelligence among individuals could be very high, possibly equal to or somewhat above that of the most intelligent individual in the historical human population.'[7] Besides genetic selection, there is now a cheap and effective tool for genetic editing—developed by Jennifer Doudna and Emmanuelle Charpentier in 2012 and called clustered regularly interspaced short palindromic repeats-associated protein 9, or CRISPR-Cas9—that could be used to delete genes and even to insert new DNA information into a cell.[8] A Chinese researcher, He Jiankui, created the world's first CRISPR baby by modifying the DNA of a human embryo and then implanting it in a woman's uterus in 2018, for which he was jailed in China for violating ethical standards.[9] The U.S. intelligence community characterized the gene editing technology as a potential weapon of mass destruction.[10]

Information and Computer Technology

Many transhumanists believe that the human brain is a biocomputer, and that a human mind could be uploaded into a computer, perhaps a computer inside a robot body, or a virtual reality.[11] Not only could a mind be uploaded, but even a society could be transferred into a computer system. For example, roboticist Hans Moravec suggested that

> a human brain equivalent can be encoded in less than one hundred million megabytes, or 10^{15} bits. If it takes a thousand times more storage to encode a body and its surrounding environment, a human with living space might consume 10^{18} bits, a large city of a million inhabitants could be efficiently stored in 10^{34} bits, and the entire existing world population would fit in 10^{36}.[12]

Some transhumanists speculate that human civilization could be transferred into a computer simulation in which everybody could live forever and live happily, somewhat like in the movie *The Matrix*.[13]

While the prospect of uploading minds seems to lie further in the future, there is the expectation that some aspects of the human mind could be replicated in a computer, leading to powerful artificial intelligence (AI) that could supplant or exceed human intelligence in many contexts. The National Commission on Artificial Intelligence stated in its final report that

> [t]he rapidly improving ability of computer systems to solve problems and to perform tasks that would otherwise require human intelligence is transforming many aspects of human life and every field of science. It will be incorporated into virtually all future technology. The entire innovation base supporting our economy and security will leverage AI. How this 'field of fields' is used—for good and for ill—will reorganize the world.[14]

It is expected that AI will be able to assist human decision-making in practically every field of human activity and endeavor. AI systems are based on a combination of big data, neural networks, and algorithms and can be trained to perform specific tasks better than humans. In theory, AI could create a world of abundance in which humans no longer need to work since enough wealth could be produced by intelligent machines to support everybody.[15] By leveraging AI, humans would become smarter, especially if their brains could be connected directly to computers to constantly receive and analyze information.

Nanotechnology

Nanotechnology is about engineering at the molecular level, which was first suggested by Richard Feynman in a 1959 lecture titled 'There Is Plenty of Room at the Bottom.' The idea was developed much further by Eric Drexler in his 1986 book *Engines of Creation*. Drexler hypothesized that self-replicating nanomachines could be built and could function similarly to biological replicators such as 'viruses, bacteria, plants, and people.' He suggested one should '[i]magine such a replicator floating in a bottle of chemicals, making copies of itself . . . [e]ach copy . . . will build yet more copies,' resulting in exponential growth. Some of the replicators will be assemblers, which are '[m]achines able to grasp and position atoms [and they] will be able to build almost anything by bonding the right atoms together.'[16] Biological nanomachines produced by nature to manage biological processes can be replicated and programmed for specific tasks. According to Sonia Contera, 'DNA nanorobots inspired by the ribosome have been developed— molecular-scale assemblers that in the future will be programmable to construct any desired molecule.'[17]

In terms of transhumanism, nanotechnology will likely be an important aspect of the technological colonization of the human body. Ray Kurzweil was one of the first to popularize the idea of injecting people with nanobots: 'Nanobots launched into our bloodstreams could supplement our natural immune system and seek out and destroy pathogens, cancer cells, arterial plague, and other disease agents. In the vision that inspired cryonics enthusiasts, diseased organs can be rebuilt.'[18] In short, nanotechnology could make it possible to grow organs tailored to individuals. The ability to design and build proteins can be used for drug and vaccine development, which was demonstrated in 2010 by David Baker, at the University of Washington, who built a protein to approximate HIV to promote the production of antibodies.[19]

According to a team of Indian researchers, '[n]anotechnology has played a significant role in the success of these vaccines. Nanoparticles (NPs) aid in improving stability by protecting the encapsulated mRNA from ribonucleases and facilitate delivery of intact mRNA to the target site.'[20] In an unprecedented effort, the COVID-19 'nanovaccines' based on lipid nanoparticles (LNPs) have been injected into over 5.5 billion people worldwide, or about 69 percent of the world population.[21] Effectively, a large portion of humanity have had their cell function altered with a nanotechnology-enabled gene therapy product.[22] This is no small step towards a transhumanist future.

Neurotechnology

The human brain is incredibly complex, with over 86 billion neurons and over 100 trillion synaptic connections. For a long time, the exact functioning of the brain was a mystery, but science has made substantial progress in decoding the brain thanks to various neural monitoring and imaging technologies such as electroencephalography (EEG), positron emission tomography (PET), magnetic resonance imaging (MRI), and functional magnetic resonance imaging (fMRI). In 2013, the Obama administration launched the BRAIN Initiative, which offered $100 million in additional research money to an effort to map the human brain and could lead to treatments for brain diseases such as Alzheimer's, epilepsy, and traumatic brain injury.[23]

There are many ways in which neurotechnology can enhance human performance—for example, through nootropic drugs and brain stimulation to improve brain function or through neural monitoring that can flag subconscious cues to a user or alert a user if they are about to fall asleep when operating a vehicle. However, the ultimate goal seems to be the development of a brain–computer interface (BCI) that can connect a brain seamlessly to a computer to receive or send information.[24] An EEG or implanted neural monitoring device would interpret or decode brain signals and translate them into text that can be transmitted wirelessly, which is an approach called 'synthetic telepathy.'[25] A user could also receive information by way of brain stimulation, which was demonstrated by researchers from the University of Washington in 2014.[26]

Elon Musk's company Neuralink, launched in 2016, is trying to develop commercially an implanted BCI that would allow people to connect themselves to the internet and to make phone calls with the help of an external device that can be controlled by thought.[27] Musk argued that BCIs would be critical for enabling humans to compete with AI. He stated:

> We're going to have the choice of either being left behind and being effectively useless or like a pet—you know, like a house cat or something—or eventually figuring out some way to be symbiotic and merge with AI.[28]

People may be compelled to accept brain implants or perhaps noninvasive BCIs as a means to remain competitive in relation to AI systems, which are expected to displace at least half of all existing jobs.[29]

The Dark Side of Transhumanism

Transhumanism has much allure since there are few people in the world who are fully content with their natural limitations and individual human flaws that could, perhaps, be overcome with the help of technology. The popularity of superhero movies attests to the widespread human desire to acquire superhuman capabilities.[30] The underlying assumption is always that any modification of humans would be an improvement that benefits individuals or that the modification would enhance an individual's physical and mental capabilities. This may not always be the case.

Matthew Liao, Anders Sandberg, and Rebecca Roache have argued that genetic engineering could be used to address broader societal issues such as climate change. They provocatively suggested genetically modifying humans to make them intolerant to eating meat (to do away with a natural desire for meat as meat production causes too many greenhouse gas emissions) and to make them smaller (smaller humans eat and consume less, and it takes less energy to transport them); using cognitive enhancement to lower birth rates (to enhance the educational attainment for women, which is negatively correlated with fertility); and using pharmacological enhancement to strengthen altruism and empathy (to make people more likely to care about climate change).[31] A cynic may think that the authors consider humans to be little more than biological machines that need to be optimized for low consumption by being turned into midgets who are unable to consume meat, who no longer have any interest in reproduction, and who are designed to be so naïve that they lack any understanding of how they may be economically exploited. In short, it is not difficult to see how transhumanism could diminish rather than enhance human abilities and the human experience for the supposed good of society.[32]

Transhumanism creates some obvious societal dilemmas as it relates to social inequality and some less obvious threats as it concerns privacy, human freedom, and the security or survival of individuals and groups. Let's start with the obvious: The human race could be split into two or more distinctive species, namely those humans who will be technologically enhanced and those humans who either cannot afford to or do not want to be enhanced or modified. Transhumanist philosopher Yuval Noah Harari argued

> One of the dangers is that we will see in the coming decades a process of . . . greater inequality than in any previous time in history because for the first time, it will be real biological inequality. If the new technologies are available only to the rich or only to people from a certain country then Homo sapiens will split into different biological castes because they really have different bodies and—and different abilities.[33]

This kind of technology-induced variance in human abilities and characteristics cannot be (in the view of Harari) overcome by any kinds of social policies, such as education or equal opportunity policies that offer social mobility to the less fortunate. The unenhanced (or maliciously modified) could thus become a permanent underclass or even a slave class.

Furthermore, there could be substantial social pressure for people to accept certain 'enhancements' that take away their privacy and, perhaps, much of their freedom. People might have to choose between accepting a modification to their bodies or never being able to earn a living.[34] Then, there is also the possibility that what is presented as transhuman enhancement is actually a hidden method for limiting life spans, reducing the ability of people to think for themselves, perhaps even denying people the ability to reproduce by themselves. Technologist Bill Joy warned two decades ago that

> [d]ue to improved techniques the elite will have greater control over the masses; and because human work will no longer be necessary the masses will be superfluous, a useless burden on the system. If the elite is ruthless they may simply decide to exterminate the mass of humanity. If they are humane

they may use propaganda or other psychological or biological techniques to reduce the birth rate until the mass of humanity becomes extinct, leaving the world to the elite.[35]

The fundamental challenge with respect to NBIC technologies is that they are generally being developed in the private sector and are incredibly difficult for governments to regulate. At present, many of these technologies are still experimental, and nobody can fully understand the inherent risks. For example, any manipulation of human DNA or even of the genome of plants and animals used for human consumption can create unforeseen problems in the long run.[36] NBIC enhancements may have security vulnerabilities that can be exploited by nefarious actors. This does not even include the possibility that people could be persuaded or tricked under a false pretext by governments or corporations into accepting body modifications that are meant to diminish their health, fertility, or longevity, or that require them to 'subscribe' to constant biotechnological interventions in order to stay alive.

Surveillance

NBIC technologies enable new possibilities for the systematic surveillance of a large population, becoming more intrusive and more comprehensive. All authoritarian and totalitarian governments use extensive surveillance over their populations in fear of revolutions or rebellions that could overthrow them. In the past, repressive governments had to employ large numbers of internal security personnel to monitor communications and recruit domestic informants in order to be able to monitor at least a significant portion of their population. In 1989, the East German secret police, nicknamed the Stasi, employed over 102,000 officers, a minimum of 174,000 NCOs, and perhaps up to 2 million unofficial informants in order to surveil a population of 17 million.[37] East Germany likely had established the most extensive surveillance system in history in terms of the overall manpower devoted to monitoring the population.[38] However, compared with current technological capabilities available to many governments around the world, the Stasi looks like amateurs. Even more concerning is that mass surveillance has gone global and extends beyond a state's territory.[39] It is also now a capability in the hands of some nonstate actors, such as multinational corporations and major IT companies.

Online Data Collection and Aggregation

Few people are even fully aware of the enormous data collection carried out by governments and corporations, capturing massive amounts of personal data, including home addresses, web surfing habits, shopping habits, movement patterns, political views, social networks, lifestyle, income and credit, hobbies, vices, and much more. Much of this data is volunteered by users, often unsuspectingly, in exchange for free online services, such as using online shopping websites or social media platforms. Other data is surreptitiously collected through online trackers (spyware).[40] Consumers have also become accustomed to the idea of 'smart devices' that constantly listen and collect data on users that is sent back to the manufacturer as a standard product feature.[41] It is all part of what Shoshana Zuboff has termed 'surveillance capitalism.'[42] IT corporations have acquired tremendous amounts of data on their users that can be sold to other corporations to be used for marketing and to governments trying to skirt domestic laws or trying to acquire data on foreign populations. The Chinese government even requires Chinese companies to share their international customer data with it, which has become a huge concern with respect to the popular social media platform TikTok.[43] The CCP has reportedly collected sensitive personal data on 80 percent of all American adults.[44]

A lot of this data can be collected openly and legally, but there is also a substantial amount of data that is acquired through clandestine or illegal methods such as hacking of organizations that keep confidential records on individuals. The number of data breaches in recent years is astounding with regard to the amount of individuals' sensitive data that has fallen into the hands of government or criminal hackers: In 2013, hackers accessed the confidential records of 3 billion Yahoo email users;[45] in 2015, Chinese hackers stole the confidential records of 22.1 million former and current U.S. government employees;[46] in 2017, Chinese military hackers stole 147 million confidential records from Equifax;[47] and, in 2019, a hacker stole over 100 million customer records from the credit card company Capital One.[48] Sooner or later, all of this data ends up on the dark web to be sold anonymously for cryptocurrency. Either way, the data is available for sale and can be acquired by both legal and criminal entities, including foreign governments, corporations, and hacker groups, to be used for a variety of purposes, many of which are malicious.

Behavioral Prediction and Influencing

The generation of vast amounts of digital personal data through online and other surveillance has driven the recent big data revolution in AI. The argument goes that any state or corporation that has more behavioral data at its disposal for training AI systems will ultimately lead in the development of more powerful AI and can produce better and more competitive AI-based products and services. According to AI entrepreneur Kai-Fu Lee, 'China's alternate digital universe now creates and captures oceans of new data about the real world. That wealth of information on users . . . will prove invaluable in the era of AI implementation.'[49] While the development of better AI incorporated into products and services seems to be to the benefit of consumers, there is also a dark side to it—namely, the possibility of using big data related to the behavior of millions, or even billions, of people for behavioral manipulation on an unprecedented scale and with unprecedented sophistication. Zuboff argued:

> [s]urveillance capitalism unilaterally claims human experience as free raw material for translation into behavioral data. Although some of these data

are applied to product or service improvement, the rest are declared as a proprietary behavioral surplus, fed into advanced manufacturing processes known as 'machine intelligence,' and fabricated into prediction products that anticipate what you will do now, soon, or later. Finally, these prediction products are traded in a new kind of marketplace that I call 'behavioral markets.'[50]

Zuboff has pointed out that

automated machine processes not only *know* our behavior but also *shape* our behavior at scale . . . the goal [of surveillance capitalism] now is to *automate us*. In this phase of surveillance capitalism's evolution, the means of production are subordinated to an increasingly complex and comprehensive 'means of behavioral modification.'[51]

Former Secretary of Homeland Security Michael Chertoff has similarly warned that massive data collection by governments and corporations poses an enormous threat to individual autonomy and freedom. He argued:

When government has total access to your information, the practical reach of its authority is almost limitless, transgressing the formal constitutional or statutory limits on official power. Just as alarming, unfettered access to that data can allow private enterprises or groups to pressure, manipulate, or incentivize personal behavior without any public accountability. As is illustrated by the increasing phenomenon of online bullying, communities with which we have no connection can use data about us to retaliate or annoy us if they don't like our political or even aesthetic opinions.[52]

The concern is that personal data can be systematically exploited to influence, manipulate, and perhaps even coerce individuals to make decisions that are not in their best interest, which can be to buy certain unneeded products or services, to vote in a certain way, or to engage in certain behaviors such as attending or not attending a protest.

The Cambridge Analytica scandal showed how the exploitation of personal data can be used for manipulating the outcome of a U.S. presidential election.[53] Media entrepreneur Steve Bannon, with money from the Mercer family, created Cambridge Analytica, which acquired data from 87 million American Facebook accounts for data mining to create psychological profiles for each user. The data and analysis were subsequently used for targeted messaging and fine-tuning the Trump campaign.[54] Major tech companies such as Google can rig elections by manipulating search results that favor one candidate over another.[55] Facebook could send an alert to voters of one party (as revealed in their profiles) to go out voting while not sending an alert to the voters of the other party, at no cost.[56] Essentially, globally operating tech companies can influence election outcomes according to their own political preferences, making a mockery of liberal democracy.

Wearables and Implants

Since the 1990s, people have become accustomed to carrying electronic devices with them in the form of cell phones, smartphones, and laptops. Many people are already addicted to smartphones, spending over 4.8 hours per day on their mobile phones using various social, photo, and video apps.[57] There is also a range of wearable devices such as smart watches, augmented reality glasses, and fitness devices, some of which also monitor body functions, location, and movement. This data can be used for tracking physiological and mental states. For example, it could be possible to determine with the data, which can be transmitted online, whether an individual is stressed or relaxed. Combined with 5G or the Internet of Things, the data produced by these wearables—and also, in the future, biosensors implanted in the body—will form an Internet of Bodies, 'which introduce[s] an even more intimate interplay between humans and gadgets.'[58] The generated data can be used in relation to how a user responds to a particular stimulus such as a message or an image to tailor the stimulus to match consumer preference, which is called neuromarketing.[59]

As early as 2013, the motion-sensing gaming input device Microsoft Kinect could detect facial expressions, measure speed movements, recognize gestures, and optically measure the heart rate of a user, requiring no wearable sensor.[60] This kind of technology could be covertly deployed and used to track the emotions of people in relation to what they see on a screen, such as their smartphone screen or computer monitor. Alternatively, corporations could surreptitiously collect users' health data. They could sell the data to insurance companies that might want to improve compliance with prescribed exercise regimes, or to anybody else who is interested in health information.[61] Wearables can also record location data, which enables the geographic tracking of users. In 2017, a global heat map for the use of fitness trackers was published on the internet and revealed the locations of several secret U.S. bases in Syria and Africa.[62]

The next step is clearly to implant devices in the body and to allow bodies and brains to interact more directly with their environment and connect to the internet via wireless 5G networks. RFID-type implants that contain individual information and that allow people to prove their identity—for example, to control access to buildings or rooms or to banking/financial services—had already been developed in the 1990s.[63] At least several thousand people in the world have already been implanted with RFID chips.[64] The argument here is mostly convenience and greater security, as a person does not need to carry ID, credit cards, or a smartphone to prove their identity or transact, which might mean that it could also be harder to steal their identity (unless the information on the chip could be cloned).[65] In the future, implants could be so small that they could be injected into the body and wirelessly transmit data on body functions to a nearby device such as a smartphone.[66] The technology exists to insert microchips inside organs, to make them very difficult, if not impossible, to remove.[67] This kind of technology could allow governments and corporations to systematically track the location, health, and mental state of billions of people in real time.

World Simulation and Digital Twins

The great availability of data on most individuals constituting the world population of, currently, 8 billion makes it theoretically possible to simulate the entire world through sophisticated models of human interactions. In 2006, researchers from Purdue University proposed a project to the U.S. Department of Defense called Synthetic Environment for Analysis and Simulation (SEAS); its aim was to create a virtual environment based on the real world for modeling and simulation, called the Sentient World Simulation (SWS). The concept paper stated:

> The ability of a synthetic model of the real world to sense, adapt, and react to real events distinguishes SWS from the traditional approach of construct- ing a simulation to illustrate a phenomena. Behaviors emerge in the SWS mirror world and are observed much as they are observed in the real world. Basing the synthetic world in theory in a manner that is unbiased to specific outcomes offers a unique environment in which to develop, test, and prove new perspectives.[68]

According to a press release, the 'Sentient World Simulation (SWS) will be a con- tinuously running, continually updated mirror model of the real world that can be used to predict and evaluate future events and courses of action.'[69] The main idea is to collect data from the internet in real time to populate a simulated version of the world and produce an accurate model of the real world that can be used to under- stand emerging trends and also to war-game how certain events or actions would affect the behavior of a given population.

The concept paper pointed out 'the unprecedented ability to train in a live and comprehensive synthetic environment that is validated by theory and up to date with the real world'; the ability to provide 'measurements of emotional arousal for a group of people' based on a combination of real-world data and simulation; the inclusion of '[p]lanning tools [that] enable plans to be developed for temporally and spatially fine-grained actions as well as long-term actions and to place com- bined plans into a single playbook'; the ability to configure the SWS by using data limited to a 'localized region of the world'; and the ability to test 'Psychological Operations (PSYOP) and Civil Affairs activities, capable of illustrating the impact of these activities on populations.'[70]

According to a brief article in *The Register* from 2007, '[b]y applying theories of economics and human psychology, its developers believe they can predict how individuals and mobs will respond to various stressors.'[71] The article also claimed that 'Homeland Security and the Defense Department are already using SEAS to simulate crises on the US mainland,' and that a simulation based on the SWS and used by the Joint Innovation and Experimentation Directorate of the U.S. Joint Forces Command

> is now capable of running real-time simulations for up to 62 nations, includ- ing Iraq, Afghanistan, and China. The simulations gobble up breaking news,

census data, economic indicators, and climactic events in the real world, along with proprietary information such as military intelligence.[72]

No recent information on the SWS can be found anywhere, which suggests that the project is highly classified.

The SWS aimed to produce what is called a 'digital twin,' which is a digital representation of a real object that also interacts with the real object through data exchanges. The idea is to use digital twins for prediction and, potentially, control of complex, dynamic systems related to infrastructure, geography, production processes, traffic, financial markets, cities, and societies.[73] Particularly relevant here is its use for the planning of 'smart cities' and 'smart societies,' which would be largely run by AI. Dirk Helbing and Javier Argota Sánchez-Vaquerizo warned that such an approach of constantly tracking individual views and behaviors as a basis for policies and regulations could threaten individual freedoms and democracy:

> Recently, it has even been proposed that considering detailed knowledge about everyone's opinions and preferences would allow one to create a democratic post-choice, post-voting society. Mass surveillance could create entirely new opportunities, here. The underlying technocracy would automatically aggregate opinions, while politicians would no anymore be needed to figure out and implement what people want and need. Rather than an upgraded democracy, however, such a society could become a novel kind of digital populism, in which the will of majorities might be relentlessly imposed on everyone, thereby undermining the protection of minorities and diversity.[74]

In other words, there is the danger that societal automation could lead to a new type of totalitarianism that, by technological means, homogenizes and normalizes society to the point that there is no longer any meaningful individual choice, as everything is collectivized. Furthermore, once such a system is put in place, there is no way for individuals to opt out or for a society to put an end to technocratic control. The authors suggest that

> a digital twin of society, which includes detailed digital twins of its people, could be easily abused. For example, by knowing the strengths and weaknesses of everyone, one could trick or manipulate everybody more effectively, or mob them with hate speech on social media . . . a digital twin of society would also make it possible to determine how much one can pressure people without triggering a revolution, or figure out how to overcome majorities, how to break the will of people, and how to impose policies on them, which do not represent their will.[75]

It is not a stretch to suggest that complex simulations using digital twins could be used to impose sustainability policies on a global population, many of which will be immensely unpopular and, hence, will require sophisticated means of persuasion.[76]

Technologies of Population Control

NBIC technologies can enable the macro control of society to the point that it makes individual free will obsolete. The end point of current trends would be Aldous Huxley's 'scientific dictatorship,' technocracy, or neofeudalism, in which a tiny elite of technocrats manages every aspect of human life and human relationships. Huxley stated: 'The nearly perfect control exercised by the government is achieved by many kinds of nearly non-violent manipulation, both physical and psychological, and by genetic standardization.'[77] Some 5GW belligerents may seek to subvert a free society by putting in place technologies that are so comprehensive and powerful that resistance by groups or individuals becomes impossible. This can be done by constructing an electronic control grid and/or by covertly using chemicals, biochemicals, biological agents, and electromagnetic weaponry to make members of a society more suggestible/obedient, to alter their mental states, degrade their cognitive abilities, and selectively eliminate individuals or groups that are likely to resist—either in the short term, through covert neutralization, including defamation, incapacitation, or assassination, or in the long term, through causing infertility in problematic/undesirable ethnic, racial, and religious groups.

Programmable Money

Few people consider money or currency to be an instrument of war, but financial warfare has been practiced for decades in the form of counterfeiting, financial sanctions, and currency manipulation.[78] The invention of Bitcoin in 2008, laid out in the famous Satoshi Nakamoto White Paper, has added a new dimension to financial warfare as it is digital money that is a bearer instrument and can be transacted peer-to-peer in a permission-less network and therefore escapes the financial control of governments.[79] For more than a decade, governments have tried to paint cryptocurrency as a technology useful solely to criminals, terrorists, and money launderers while quietly working on new digital money that could replace the current monetary system. The new digital monetary system would be global and would be designed by a democratically unaccountable and little-known international organization called the Bank for International Settlements (BIS), based in Basel.[80] Few recent technological developments could have more serious implications for individual freedom than central bank digital currencies, or CBDCs.

In June 2021 the BIS published a report titled 'CBDCs: An Opportunity for the Monetary System.' The CBDCs would be based on blockchain technology, referred to in the report as distributed ledger technology (DLT), and would share some similarities with Bitcoin and its derivatives, but would in other aspects be very different from it. Essentially, the CBDCs would be run as a private blockchain with vetted validators of transactions, which is a permissioned system.[81] The BIS report suggested that CBDCs could be a two-tier system, with the central banks providing the foundations, and private payment service providers (banks) managing the CBDCs wallets on behalf of customers.[82] The BIS predicted that, within three years (i.e. by 2024), about 20 percent of the world population will use CBDCs.[83]

The Federal Reserve Bank of Boston has begun working on Project Hamilton, which is development project for a CBDC and was announced in February 2022.[84] The main advantage of digital money is that it is programmable, in the sense that various conditions can be included as to when and how funds can be spent, which gives the issuer of the currency complete visibility and more control over all transactions taking place in an economy.[85] Citizens could have a government digital wallet on their mobile phone to directly receive universal basic income payments from the government. The central banks could put expiry dates on funds, implement negative interest rates to stimulate the economy or to manage inflation, or enforce social policies.[86] To make the system work globally, a digital ID linked to a unique individual electronic wallet would be needed for every human being on Earth.[87]

It is not difficult to understand how CBDCs could be abused to establish a system of unprecedented economic, political, and social control. If a CBDC had an expiry date, it would become impossible for average people to accumulate wealth or even to pass on wealth to their children, as all money would have to be spent before it expires. A government could make an individual's ability to spend money conditional on having a good social credit score or on having sufficient carbon credits. In China, social credit has already been implemented. According to Paul Scharre,

> [i]ndividuals can have points subtracted from their score or be blacklisted for eating or playing loud music on the train, running a red light, failing to show up to a restaurant reservation, not leashing one's dog in public, not properly sorting trash and recycling, and (of course) jaywalking.[88]

With CBDCs, it would be easy to put in place individualized restrictions, where specific individuals could be restricted from buying certain goods or services—for example, to limit their consumption or movement. It would be easy to freeze wallets or remove funds for any reason, leaving individuals with no money to buy food or make essential payments.[89] Fines could be automatically withdrawn from wallets when an individual breaks a rule or misbehaves in the eyes of the authorities, with perhaps no recourse available.[90] If there was no alternative system for financial transactions, people could be quietly removed from society and forced to starve by just switching off their wallets. If the new digital monetary system was global (as in a unified ledger envisioned by the BIS), there would be nowhere to go for individuals who are blacklisted by a central bank. If the wallet was linked to an implanted chip, people could be coerced by an unaccountable international organization to accept implants in order to participate in the economy.[91] The only thing that could stand in the way of the implementation of a CBDC is populations' distrust.[92]

The New Biowarfare Threat

Historical examples of biological warfare are rare. According to Gregory Koblentz, 'biological weapons have never been used on the battlefield and their development has always been conducted under the strictest secrecy.'[93] There are only a handful of nations that are known to have had large-scale biowarfare programs producing and stockpiling biological munitions, namely Japan, the Soviet Union, the United Kingdom, the United States, Iraq, and North Korea.[94] Currently, the U.S. intelligence community estimates that only North Korea has an active offensive biowarfare program with the 'capability to produce sufficient quantities of biological agents for military purposes upon leadership demand.'[95]

Biological weapons are useless as deterrents owing to the need to achieve surprise, which necessitates secrecy. They are also useless for retaliation as the effects would be delayed and would be uncontrollable.[96] This means that biowarfare capabilities are only suited for offensive use against an unsuspecting or unprepared target, most likely in the form of a covert or bioterrorism attack rather than an overt military attack with biological munitions. The U.S. military and intelligence extensively evaluated the threat of a covert biological attack on the United States in the 1950s and 1960s and concluded that the United States was vulnerable to an enemy covertly spraying biological agents from aircraft, from ships near the coast, from rooftops, or in subways in major cities.[97]

According to former KGB officer Alexander Kouzminov, the Soviet Politburo had plans to covertly disperse biological agents in the West using 'illegals' (deep cover operatives) in the event of war, targeting 'public drinking water supplies, food stores, and processing plants; water purification systems; vaccine, drug and toxin repositories; and pharmaceutical and biotechnological plants.'[98] The KGB's Department 12, in charge of the procurement and development of biological and toxin weapons, was paired with Department 8, which was in charge of 'planning and carrying out acts of terrorism and sabotage and launching diversionary missions in target countries in the event of war.'[99] Stanislav Lunev also suggested that it was Russian military doctrine to poison the water supplies of large cities with chemical and biological agents.[100] The Soviet Union is now gone, but the threat of a covert biological attack has not disappeared, and the impact of the COVID-19 pandemic has heightened states' interest in biowarfare.[101]

Fifth generation warfare belligerents are likely to covertly use biological agents or carry out bioterrorism attacks, which can blend state and nonstate capabilities. Thomas Hammes argued that '[t]he October 2001 anthrax attack on Capitol Hill may have been the first 5GW attack,' and that '[t]he anthrax attack provided stark evidence that today a single individual can attack a nation state.'[102] The Amerithrax attack was enabled through access to anthrax from a U.S. defense lab and expertise gained from a state biodefense program that enabled the weaponization of a deadly biological agent.[103] Similarly, Thomas Barnett has pointed out, bioterrorism might target food and water with the objective of destabilizing a society. He wrote, 'what I expect Fifth Generation Warfare . . . to be all about in the twenty-first century: biological terror to create economic dislocation and loss (along with the usual panics).'[104]

It has been argued that nonstate actors could benefit from the declining cost and relative ease of using modern biotechnology for the development and/or production of biological agents.[105] However, bioterrorism is still exceedingly rare. According to a recent study covering a 50-year period (1970–2019), there have only been 33 bioterrorism incidents, which resulted in nine deaths and 806 injuries (almost all of the injuries related to just two incidents involving salmonella in food).[106] One issue could be that bioterrorism attacks can go unnoticed if they fail to produce visible effects. For example, Aum Shinrikyo carried out at least nine biological attacks in Japan, directed at the parliament, the Imperial Palace, and the U.S. naval base at Yokosuka, which were not detected at the time.[107] Fifth generation warfare belligerents with access to state-like capabilities may choose to carry out covert biological attacks with pathogens that are natural or could have emerged naturally and have a low lethality, but may incapacitate or debilitate victims or cause infertility.

Genetic Bioweapons

Ken Alibek revealed, after his defection, that the Soviet biowarfare program was working on a genetically engineered chimera virus using recombinant DNA that combined smallpox and Venezuelan equine encephalitis (a virus that attacks the brain) to produce 'a superweapon capable of triggering both diseases at once.'[108] Using genetic engineering techniques, the Soviet Union also worked on antibiotics-resistant strains of bacterial diseases such as anthrax and the plague.[109]

The now defunct think tank the Project for a New American Century argued, in 2000, that 'advanced forms of biological warfare that can "target" specific genotypes may transform biological warfare from the realm of terror to a politically useful tool.'[110] According to a National Academy of Sciences report,

> [b]iotechnology advances may offer new opportunities for a malicious actor to influence the overall impact of an attack or the specific individuals affected, such that an agent could be deployed over a broad geographic area but only sicken targeted individuals.[111]

A report by Cambridge University's Centre for Existential Risk claimed that '[m]ore nefarious hands could (as they have before) develop pathogens and toxins to spread through air, food and water sources . . . a bio-weapon could be built to target a specific ethnic group based on its genomic profile.'[112] As early as the 1980s, South Africa was exploring the possibility of biological or chemical agents that would be only effective against black people and/or that could cause sterility.[113] South Africa was greatly concerned about the high fertility of black people and developed a vaccine that 'could be given without the knowledge of the recipient . . . either orally or as an injection,' according to a Project Coast insider.[114]

The general idea of an ethnic bioweapon is that it targets an ethnic group in which a particular genetic marker is common as opposed to an ethnic group in which it is less common. What is needed is a sufficient genetic sample of the target population. Reportedly, China has been systematically collecting DNA data from Americans and may have exploited the pandemic to collect tens of millions of U.S. DNA samples through the Chinese company BGI that sold COVID-19 testing to the United States.[115] Paul Dabbar has claimed that '[a] Chinese general who was head of the National Defense University in Beijing publicly declared an interest in using gene sequencing and editing to develop pathogenic bioweapons that would target specific ethnic groups.'[116] The U.S. State Department has suggested that China is 'engaged in activities with dual-use applications, which raise concerns regarding its compliance with Article I of the BWC,' which means that China may be secretly working on new biological weapons.[117]

According to Filippa Lentzos, genetic engineering creates novel risks and utilizes novel delivery mechanisms 'through the enhancement of pathogens to make them more dangerous; the modification of low-risk pathogens to become high-impact; the engineering of entirely new pathogens; or even the re-creation of extinct, high-impact pathogens like the variola virus that causes smallpox,' which could be deployed with the 'use of drones, nano-robots, even insects.'[118] Lentzos further suggested that new biotechnology approaches could circumvent the BWC and 'instead of using bacteria or viruses to make us sick, directly target the immune, nervous or endocrine systems, the microbiome, or even the genome by interfering with, or manipulating, biological processes.'[119] Similarly, David Malet pointed out that

> subsequent innovations in genetic engineering offer novel approaches through the disruption of the human body's basic functions by rewriting the target's genetic code. Although these new 'direct effect' weapons would be delivered by vectors (engineered synthetic viruses), they would effectively turn the body's regulatory systems against itself rather than using foreign substances.[120]

Harmful genetic information that tricks a cell into producing a pathogen inside it could be delivered by way of genetically modified mosquitoes.[121] It might be difficult to determine that a biological attack had occurred since the virus that delivers the harmful genetic information could be innocuous. It is also possible to imagine that the insertion of genes through a virus could be used to genetically modify members of a population without their knowledge or consent, perhaps to limit their cognitive abilities, their physical capabilities, or their overall life expectancy.

Covert DEW Attacks as Neurowarfare

Another possibility for conducting a covert attack on a population, a subgroup, or even specific individuals could be the use of nanotechnology, possibly in combination with directed energy in order to surreptitiously alter mental states, emotions, and perceptions and influence behavior. Russian military analysts have described antipersonnel radiofrequency (RF) weapons that they refer to as 'psychotronic weapons,' 'weapons based on new physical principles,' or 'information weapons.' For example, Vladimir Belous discussed

[t]he study of the impact of bio-energy on command and control and communication systems, and electronic equipment, as well as the development of artificial energy generators to impact on enemy personnel and noncombatants with the aim of inducing an abnormal psychic state.[122]

According to Russian sources, psychotronic weapons are capable of interfering with the 'central nervous system and mentality of people, with the aim of amplifying fear and paralyzing the will.'[123] Russian military theorists S.G. Chekinov and S.A. Bogdanov alluded to weapons with 'psychotronic effects' that can undermine 'the capabilities of the adversary's troops and civilian population to resist.'[124] Timothy Thomas found further references to psychotronic weapons/information weapons that are non-lethal, psychological weapons in numerous Russian military articles written between 2000 and 2020.[125] In 2012, Putin alleged the existence of electromagnetic 'high-tech weapons systems [that] will be comparable in effect to nuclear weapons, but will be more acceptable in terms of political and military ideology.'[126]

Robert McCreight calls these weapons 'neurostrike weapons.' He suggests that such a weapon 'entails a hand held, or platform mounted, mixture of an RF, directed energy pulse or *neurocognitive disrupter, combined with acoustic wave dynamics* which is designed to harm, disable or permanently damage a human brain.'[127] A recent report by the think tank CCP Biothreat Initiative has claimed that the CCP is working on neurostrike weapons, as defined by McCreight, in support of psychological warfare in the Indo-Pacific theater. It suggested: 'Their [the CCP's] new landscape of NeuroStrike development includes using massively distributed human–computer interfaces to control entire populations as well as a range of weapons designed to cause cognitive damage.'[128]

There are a couple of plausible scenarios in which psychotronic weapons or neurostrike weapons, as defined by McCreight, could be used in covert attacks against civilians during peacetime in order to achieve a strategic objective. In short, they could be used offensively or defensively, or either to destabilize a political system or to stabilize it. If it were possible to use DEWs (electromagnetic or acoustic) to affect emotional states, one could create an event, such as the assassination of a popular leader or the murder of an innocent civilian by a police officer, that has a high potential to cause public outrage, which could then be amplified by technical means to cause societal chaos.[129]

For example, the Rwandan president was assassinated by unknown assassins in April 1994, which sparked the government-encouraged genocide of the Tutsi minority by the Hutu majority in a bid to quickly win the civil war and eliminate the Tutsi threat. Between 507,000 and 850,000 Tutsi and about 30,000 Hutus were killed.[130] Mark Safranski has pointed out that 'the Rwandan genocide is notable for the recruitment of an enormous number of participants, where every Hutu was expected to play the role of an enthusiastic SS man.'[131] According to a popular theory, the genocide was driven, or at least greatly encouraged, by inflammatory radio broadcasts that framed the conflict in ethnic terms and increased fear of the out-group.[132] In other words, if it is possible, by way of technical manipulation of the emotions of one societal group, to induce extreme fear of another societal group that is visibly distinct, one can trigger a civil war or genocide. One only needs to provide opportunity and capability (in this case, security personnel standing down and machetes for every Hutu).[133]

A defensive use of psychotronic weapons could be covertly attacking political opponents or dissidents as a means of neutralizing them, as in rendering them incapable of any effective resistance to a government. Unlike overt measures of repression such as incarceration or assassination, covert DEW attacks would not result in any embarrassment or criticism of a government that may try to present itself to the world as democratic and legitimate. The best-known example of a state practicing repression that covertly attacked and neutralized dissidents, some of whom were internationally recognized, was East Germany. The Stasi used a technique it termed *Zersetzung* (decomposition or disintegration), which targeted at least 4,000–5,000 individuals per year.[134] The Stasi defined *Zersetzung* as

> an operative method to effectively combat subversive activity, in particular during the process of handling the subject. Z(ersetzung) entails the use of various political and operational procedures to exert influence over hostile or negative persons, in particular over their hostile ideas and convictions, in such a way that the latter can be shaken or gradually altered, or contradictions and differences between hostile forces can be aroused, utilized or strengthened.[135]

The techniques used included psychological torture and were carried out by informal Stasi collaborators. They frequently resulted in the victims developing serious health and mental health problems.[136] The Stasi also planted radioactive materials in the apartments and clothing of victims to cause health damage and induced cancer by using X-ray machines to covertly irradiate inmates in political prisons.[137] Psychotronic or neurostrike weapons could damage the brains of dissidents from a distance and covertly, leaving behind little evidence. They are perfect weapons for covert repression.

Conclusion

Transhumanism is an ideology that promises individuals salvation in the form of greater mental and physical capabilities and greater longevity (if not eternal life). The obvious problem is that, in the eyes of global technocratic elites, the human population is already too large to be sustainable, as resources are finite and are getting depleted at a growing rate.[138] In addition, AI and other technologies that enable more automation are making a large portion of workers superfluous in the so-called 'fourth industrial revolution.'[139] Making all people alive today healthier and live longer would exacerbate the perceived problems, rather than solve them. Large populations divided into competing nation states are also more difficult to govern. It is not hard to see why global elites would prefer a much smaller world population under a world government. The technological means to achieve a more dramatic reduction in the world population are increasingly available. Human beings could be re-engineered and programmed to be infertile, have shorter life spans, and be happy with less. As forecast by Aldous Huxley in *Brave New World*, all reproduction could be controlled by the government through in vitro fertilization of embryos grown in artificial wombs according to specifications that reflect government or societal needs. The path towards such a nightmarish future can be described as a fifth generation war waged by a tiny elite class against the '99 percent.' The 'war' is the covert, deceptive, and gradual deployment of technologies of social control, which are meant to make all potential resistance to unpopular and much more repressive societal changes impossible.

Notes

1 Stuart A. Newman, 'The Transhumanism Bubble,' *Capitalism Nature Socialism* 21, no, 2 (2010): 29.
2 Julian Huxley, 'Transhumanism,' *Journal of Humanistic Psychology* 8, no. 1 (1968): 74, 76.
3 Ray Kurzweil, *The Age of Spiritual Machines* (New York: Penguin, 2000), 13.
4 Nick Bostrom, 'Human Genetic Enhancements: A Transhumanist Perspective,' *The Journal of Value Inquiry* 37 (December 2003): 498.
5 Bostrom, 'Human Genetic Enhancements,' 499.
6 Patrick Wood, *The Evil Twins of Technocracy and Transhumanism* (Meza, AZ: Coherent, 2022), 20.
7 Nick Bostrom, *Superintelligence: Paths, Dangers, Strategies* (Oxford: Oxford University Press, 2016), 47.
8 Fazale R. Rana with Kenneth R. Samples, *Humans 2.0: Scientific, Philosophical, and Theological Perspectives on Transhumanism* (Covina, CA: RTB Press, 2019), 29.
9 Antonio Regalado, 'The Creator of the CRISPR Babies Has Been Released from a Chinese Prison,' *MIT Technology Review*, April 4, 2022, available at: www.technologyreview.com/2022/04/04/1048829/he-jiankui-prison-free-crispr-babies/
10 James Clapper, 'Statement for the Record: Worldwide Threat Assessment of the US Intelligence Community,' Senate Armed Services Committee, February 9, 2016, available at: www.dni.gov/files/documents/SASC_Unclassified_2016_ATA_SFR_FINAL.pdf, 9.
11 Kurzweil, *The Age of Spiritual Machines*, 203.
12 Hans Moravec, *Robot: Mere Machine to Transcendent Mind* (Oxford: Oxford University Press, 1999), 166–167.
13 Adam Piore, 'The Neuroscientist Who Wants to Upload Humanity to a Computer,' *Popular Science*, May 16, 2014, available at: www.popsci.com/article/science/neuroscientist-who-wants-upload-humanity-computer/
14 National Security Commission on Artificial Intelligence, Final Report, March 2021, available at: www.nscai.gov/wp-content/uploads/2021/03/Full-Report-Digital-1.pdf, 20.
15 Matt Rosoff, 'A.I. May Replace Most of Today's Jobs, but This Start-Up Investor Sees Global GDP Increasing by 50 Percent a Year within Decades,' *CNBC*, February 26, 2019, available at: www.cnbc.com/2019/02/26/sam-altman-on-ai-jobs-may-go-away-but-massive-abundance-likely.html
16 Eric Drexler, *Engines of Creation: The Coming Era of Nanotechnology* (New York: Anchor Books, 1987), 54, 58.
17 Sonia Contera, *Nano Comes to Life: How Nanotechnology Is Transforming Medicine and the Future of Biology* (Princeton, NJ: Princeton University Press, 2019), 114.
18 Kurzweil, *The Age of Spiritual Machines*, 224.
19 Contera, *Nano Comes to Life*, 113.

20 Amit Khurana, Prince Allawahdi, Isha Khurana, Sachin Allwahdi, Ralf Weiskirchen, Anil Kumar Banothu, Deepak Chhabra, Kamaldeep Joshi, and Kala Kumar Bharani, 'Role of Nanotechnology behind the Success of mRNA Vaccines for COVID-19,' *Nano Today* 38 (2021): 1.
21 Josh Holder, 'Tracking Coronavirus Vaccinations around the World,' *New York Times*, March 13, 2023, available at: www.nytimes.com/interactive/2021/world/covid-vaccinations-tracker.html
22 Helene Banoun, 'mRNA: Vaccine or Gene Therapy? The Safety Regulatory Issues,' *International Journal of Molecular Sciences* 24, no. 13 (2023), available at: https://doi.org/10.3390/ijms241310514
23 *New York Times*, 'Obama to Unveil Initiative to Map the Human Brain,' *New York Times*, April 2, 2013.
24 Samantha Masunaga, 'A Quick Guide to Elon Musk's New Brain-Implant Company, Neuralink,' *Los Angeles Times*, April 21, 2017.
25 Eric Bland, 'Army Developing "Synthetic Telepathy,"' *NBC News*, October 13, 2008, available at: www.nbcnews.com/id/wbna27162401
26 Rajesh P.N. Rao, Andrea Stocco, Matthew Bryan, Devaparatim Sarma, Tiffany M. Youngquist, Joseph Wu, and Chantal S. Prat, 'A Direct Brain-to-Brain Interface in Humans,' *PLOS One* 9, no. 11 (2014): 1–12.
27 Victor Tangermann, 'Elon Musk Says Neuralink's First Product Will Control Smartphone with Brain Implant,' *Futurism*, April 9, 2021, available at: https://futurism.com/elon-musk-neuralink-first-product-control-smartphone-brain-implant
28 Masunaga, 'A Quick Guide to Elon Musk's New Brain-Implant Company, Neuralink.'
29 Working Paper, September 17, 2013, available at: www.oxfordmartin.ox.ac.uk/downloads/academic/future-of-employment.pdf, 41.
30 Rana with Samples, *Humans 2.0*, 12.
31 S. Matthew Liao, Anders Sandberg, and Rebecca Roache, 'Human Engineering and Climate Change,' *Ethics, Policy, and Environment* 15, no. 2 (2012): 208–211.
32 Matthew Liao has argued against the deliberate diminishing of human capacities through genetic engineering: '[I]t is not permissible to deliberately create an offspring that will not have all the fundamental capacities' and they include 'the capacity to act, to move, to reproduce, to think, to be motivated, to have emotions, to interact with others and the environment, and to be moral.' Even if such a principle would be embraced by all governments in the world, who could police compliance? Rebecca Beyer, 'Is Gene Editing Ethical? It Depends,' *New York University News*, February 26, 2019, available at: www.nyu.edu/about/news-publications/news/2019/february/is-gene-editing-ethical—it-depends.html
33 Anderson Cooper, 'Yuval Noah Harari on the Power of Data, Artificial Intelligence and the Future of the Human Race,' *CBS News*, October 31, 2021, available at: www.cbsnews.com/news/yuval-noah-harari-sapiens-60-minutes-2021-10-31/
34 This possibility was actually discussed by Nita Farhani in her presentation at the World Economic Forum, January 20, 2023, available at: www.youtube.com/watch?v=VKg4NU3Pi2Y
35 Bill Joy, 'Why the Future Does Not Need Us,' *Wired Magazine* 8.04 (2000), available at: www.cc.gatech.edu/computing/nano/documents/Joy - Why the Future Doesn't Need Us.pdf
36 Ari LeVaux, 'Genetically Modified Food May Pose a Real Danger to Health,' in: Noel Merino (ed.), *Genetically Modified Food* (Farmington Hills, MI: Greenhaven Press, 2014), 111–118.
37 John O. Koehler, *Stasi: The Untold Story of the East German Secret Police* (Boulder, CO: Westview Press, 1999), 8.
38 Andreas Lichter, Max Löffler, and Sebastian Siegloch, 'The Long-Term Costs of Government Surveillance: Insights from Stasi Spying in East Germany,' *Journal of the European Economic Association* 19, no. 2 (2021): 775.

39 Tom Engelhardt and Glenn Greenwald, *Shadow Government: Surveillance, Secret Wars, and a Global Security State in a Single Superpower World* (Chicago, IL: Haymarket Books, 2014).

40 Bruce Schneier, *Click Here to Kill Everybody: Security and Survival in a Hyperconnected World* (New York: W.W. Norton, 2018), 64–68.

41 Amazon's Alexa never sleeps and constantly listens to sounds in its vicinity, which eliminates privacy as the device may record, store, and utilize the data. Grant Clauser, 'Amazon's Alexa Never Stops Listening to You: Should You Worry?' *New York Times*, August 8, 2019.

42 Shoshana Zuboff, *The Age of Surveillance Capitalism: The Fight for the Future at the New Frontier of Power* (London: Profile Books, 2019).

43 Jack Nicas, Mike Isaac, and Ana Swanson, 'U.S. Opens Review into China's TikTok,' *New York Times*, November 2, 2019.

44 Jon Wertheim, 'China's Push to Control Americans' Health Care Future,' *CBS News*, January 31. 2021, available at: www.cbsnews.com/news/biodata-dna-china-collection-60-minutes-2021-01-31/

45 Nicole Perlroth, 'All 3 Billion Yahoo Accounts Were Affected by 2013 Attack,' *New York Times*, October 3, 2017.

46 Ellen Nakashima, 'Hacks of OPM Databases Compromised 22.1 Million People, Federal Authorities Say,' *Washington Post*, July 9, 2015.

47 Katie Brenner, 'U.S. Charges Chinese Military Officers in 2017 Equifax Hacking,' *New York Times*, February 10, 2020.

48 Kate Conger, 'Ex-Amazon Worker Convicted in Capital One Hacking,' *New York Times*, June 17, 2022.

49 Kai-Fu Lee, *AI Superpowers: China, Silicon Valley, and the New World Order* (Boston: Houghton Mifflin Harcourt, 2018), 17.

50 Zuboff, *The Age of Surveillance Capitalism*, 8.

51 Zuboff, *The Age of Surveillance Capitalism*, 8.

52 Michael Chertoff, *Exploiting Data: Reclaiming Our Cyber Security in the Digital Age* (New York: Atlantic Monthly Press, 2018), 23.

53 Nic Fildes, David Bond, and Aliya Ram, 'Cambridge Analytica under Fire as Scandal Grows,' *Financial Times*, March 20, 2018.

54 Christopher Wylie, *Mindf*ck: Cambridge Analytica and the Plot to Break America* (New York: Random House, 2019), 92, 112–132.

55 Robert Epstein, 'How Google Could Rig the 2016 Election,' *Politico*, August 19, 2015, available at: www.politico.com/magazine/story/2015/08/how-google-could-rig-the-2016-election-121548/

56 Zoe Corbyn, 'Facebook Experiment Boosts Voter Turnout,' *Nature*, September 12, 2012, available at: www.nature.com/articles/nature.2012.11401

57 Jane Wakefield, 'People Devote Third of Waking Time to Mobile Apps,' *BBC News*, January 12, 2022, available at: www.bbc.com/news/technology-59952557

58 M. Gardner, 'The Internet of Bodies Will Change Everything, for Better or Worse,' *RAND*, October 29, 2020, available at: www.rand.org/blog/articles/2020/10/the-internet-of-bodies-will-change-everything-for-better-or-worse.html

59 Ferdousi Sabera Rawnaque, Khandoker Mahmudur Rahman, Syed Ferhat Anwar, Ravi Vaidyanathan, Tom Chau, Farhana Sarker, and Khondaker Abdullah Al Mamun, 'Technological Advancements and Opportunities in Neuromarketing: A Systematic Review,' *Brain Inform* 7, no. 1 (2020): 2.

60 David Pierce, 'The All-Seeing Kinect: Tracking My Face, Arms, Body, and Heart on the Xbox One,' *The Verge*, May 21, 2013, available at: www.theverge.com/2013/5/21/4353232/kinect-xbox-one-hands-on

61 Zuboff, *The Age of Surveillance Capitalism*, 215.

62 Liz Sly, 'U.S. Soldiers Are Revealing Sensitive and Dangerous Information by Jogging,' *Washington Post*, January 28, 2018.

63 Katherine Albright and Liz McIntyre, *Spychips: How Major Corporations and Government Plan to Track Your Every Move with RFID* (Nashville, TN: Nelson Current, 2005), 179.
64 Haley Weiss, 'Why You Are Probably Getting a Microchip Implant Someday,' *The Atlantic*, September 21, 2018, available at: www.theatlantic.com/technology/archive/2018/09/how-i-learned-to-stop-worrying-and-love-the-microchip/570946/
65 John Halamka, Ari Juels, Adam Stubblefield, and Jonathan Westhues, 'The Security Implications of VeriChip Cloning,' *Journal of the American Medical Informatics Association* 13, no. 6 (2006): 601–607.
66 Charlotte Edwards, 'Injectable Body Sensors Could Stream Medical Data to Smartphone,' *Medical Device Network*, March 19, 2018, available at: www.medicaldevice-network.com/news/injectable-body-sensors-stream-medical-data-smartphones/
67 Albright and McIntyre, *Spychips*, 186–187.
68 Tony Cerri and Alok Chaturvedi, 'A Sentient World Simulation (SWS): A Continuous Running Model of the Real World/A Concept Paper for Comments,' Purdue University, August 22, 2006, available at: https://business.purdue.edu/academics/MIS/workshop/papers/AC2_100606.pdf
69 Alok Chaturvedi, 'Computational Challenges for a Sentient World Simulation,' Purdue University, March 10, 2006, available at: https://archive.ph/20060911223310/http://www.purdue.edu/acsl/abstract/march10_06.html - selection-209.1–209.193
70 Cerri and Chaturvedi, 'A Sentient World Simulation (SWS),' 8–9.
71 Mark Baard, 'Sentient World: War Games on the Grandest Scale,' *The Register*, June 23, 2007, available at: www.theregister.com/2007/06/23/sentient_worlds/
72 Baard, 'Sentient World.'
73 Dirk Helbing and Javier Argota Sánchez-Vaquerizo, 'Digital Twins: Potentials, Ethical Issues, and Limitations,' in: Andrej Zwitter and Oskar Gstrein, *Handbook on the Politics and Governance of Big Data and Artificial Intelligence* (Cheltenham, UK: Edward Elgar, 2023), available at: https://papers.ssrn.com/sol3/papers.cfm?abstract_id=4167963, 1–2.
74 Helbing and Sánchez-Vaquerizo, 'Digital Twins,' 10.
75 Helbing and Sánchez-Vaquerizo, 'Digital Twins,' 11.
76 A study by the University of Leeds offers some good insights into what the targets are across several consumption categories. The ambitious goals for 2030 are: 20% reduction in new building construction, zero meat and dairy consumption, three items of clothing per person per year, zero private vehicles, one short-haul flight per person every three years, and an average lifetime for electronics of seven years. It would be incomprehensible for a majority to willingly embrace such a drastic reduction in living quality. University of Leeds, C40, and Arup, 'The Future of Urban Consumption in a 1.5°C World: C40 Cities Headline Report,' June 2019, available at: www.c40.org/wp-content/uploads/2021/08/2270_C40_CBE_MainReport_250719.original.pdf
77 Aldous Huxley, *Brave New World Revisited* (New York: Harper Perennial, 2006), 3.
78 Ricardo A. Crespo, 'Currency Warfare and Cyber Warfare: The Emerging Currency Battlefield of the 21st Century,' *Comparative Strategy* 37, no. 3 (2018): 236–237.
79 Paul Vigna and Michael J. Casey, *The Age of Cryptocurrency: How Bitcoin and Digital Money Are Challenging the Global Economic Order* (New York: St. Martin's Press, 2015), 41–42.
80 Bank for International Settlements, 'BIS Innovation Hub Work on Central Bank Digital Currency (CBDC),' available at: www.bis.org/about/bisih/topics/cbdc.htm, accessed July 20, 2023.
81 Bank for International Settlements, 'CBDCs: An Opportunity for the Monetary System,' available at: www.bis.org/publ/arpdf/ar2021e3.pdf, accessed on April 22, 2022.

82 Bank for International Settlements, 'CBDCs: an Opportunity for the Monetary System,' 81.
83 Hyunyun Jung, 'Blockchain Implementation Method for the Interoperability between CBDCs,' *Future Internet* 13, no. 5 (2021): 2.
84 Federal Reserve Bank of Boston, 'Project Hamilton Phase 1: A High Performance Payment Processing System Designed for Central Bank Digital Currencies,' February 3, 2022, available at: www.bostonfed.org/publications/one-time-pubs/project-hamilton-phase-1-executive-summary.aspx, accessed April 22, 2022.
85 Bank for International Settlements, 'III. Blueprint for the Future Monetary System: Improving the Old, Enabling the New,' *BIS Annual Economic Report*, June 2023, 98.
86 Andrew Moran, 'CBDCs with Expiration Dates, Restrictions Could Target Social Policies, Economist Tells WEF,' *Epoch Times*, June 30, 2023, available at: www.theepochtimes.com/expiry-dates-restrictions-benefits-of-central-bank-digital-currencies-economist-tells-wef_5366056.html
87 The UN is trying to get a global digital ID system in place for every person in the world through the WHO in the context of a global vaccine passport that would be needed for international travel. World Health Organization, 'The European Commission and WHO Launch Landmark Digital Health Initiative to Strengthen Global Health Security,' Press Release, June 5, 2023, available at: www.who.int/news/item/05-06-2023-the-european-commission-and-who-launch-landmark-digital-health-initiative-to-strengthen-global-health-security
88 Paul Scharre, *Four Battlegrounds: Power in the Age of Artificial Intelligence* (New York: W.W. Norton, 2023), 99.
89 The Brazilian government released the source code for the Brazilian CBDC, which revealed that it gives the Brazilian government the ability to freeze or withdraw funds from individual wallets. Pedro Solimano, 'Brazilian CBDC Allows Government to Freeze Funds, Developer Finds,' Yahoo Finance, July 16, 2023, available at: https://finance.yahoo.com/news/brazilian-cbdc-allows-government-freeze-174630324.html
90 During the Canadian trucker strike of early 2022, the Canadian government froze the bank accounts of individuals whom they deemed to be engaging in or supporting unlawful protests. CBDCs could make it much easier and quicker for a government to implement account freezes for a larger number of people. Miller Whitehouse-Levine, 'CBDCs Will Lead to Absolute Government Control,' *Cointelegraph*, April 6, 2023, available at: https://cointelegraph.com/news/desantis-is-right-cbdcs-will-lead-to-absolute-government-control
91 This possibility has been warned of by Professor Richard Werner, who is an economist at the University of Winchester and an expert in central banks. See Richard Werner, 'Crucial CBDCs with Professor Richard Werner: YOUR Future Is Being Decided,' available at: www.youtube.com/watch?v=KGpQLbZXKME, accessed on July 20, 2023.
92 The Nigerian CBDC eNaira trial failed spectacularly in late 2022 as only 0.5% of Nigerians were willing to use it. Alys Key, 'Nigeria Limits Cash Withdrawals to $45 per Day in CBDC, Digital Banking Push,' Yahoo Finance, December 7, 2022, available at: https://finance.yahoo.com/news/nigeria-limits-cash-withdrawals-45-100611213.html
93 Gregory Koblentz, *Living Weapons: Biological Warfare and International Security* (Ithaca, NY: Cornell University, 2009), 4.
94 Some more countries were known to have had smaller biological weapons that did not produce biological munitions, namely South Africa, Israel, China, France, and Rhodesia.
95 U.S. Department of State, '2021 Adherence to and Compliance with Arms Control, Nonproliferation, and Disarmament Agreements and Commitments,' April 2021, available at: www.state.gov/wp-content/uploads/2021/04/2021-Adherence-to-and-Compliance-With-Arms-Control-Nonproliferation-and-Disarmament-Agreements-and-Commitments.pdf, 49.

96 Koblentz, *Living Weapons*, 39–43.
97 Leonard A. Cole, *Clouds of Secrecy: The Army's Germ Warfare Tests over Populated Areas* (Totowa, NJ: Rowman & Littlefield, 1988), 59–69.
98 Alexander Kouzminov, *Biological Espionage: Special Operations of the Soviet and Russian Intelligence Services in the West* (London: Greenhill Books, 2005), 35.
99 Kouzminov, *Biological Espionage*, 35–36.
100 Stanislav Lunev, *Through the Eyes of the Enemy: Russia's Highest Ranking Military Defector Reveals Why Russia Is More Dangerous Than Ever* (Washington, DC: Regnery, 1998), 29–30.
101 Glenn Cross, 'Biological Weapons in the "Shadow War,"' *War on the Rocks*, November 9, 2021, available at: https://warontherocks.com/2021/11/biological-weapons-in-the-shadow-war/
102 Thomas X. Hammes, 'Fourth Generation Warfare Evolves, Fifth Emerges,' *Military Review* 87, no. 3 (2007): 21.
103 A study by three biologists published in the journal *Bioterrorism & Biodefense* argued that the FBI's conclusions were faulty, and that the anthrax did not come from Fort Detrick or involve any individual working there. Furthermore, they argued that the anthrax had been weaponized by microencapsulating the spores with silica, which would have required 'a laboratory with specialized capabilities and expertise not found at USAMRIID.' In short, the attack was much more sophisticated than publicly claimed and involved government capabilities to which an insider or insider group needed to have had access. Martin E. Hugh-Jones, Barbara Hatch Rosenberg, and Stuart Jacobsen, 'The 2001 Anthrax Attack: Key Observations,' *Bioterrorism & Biodefense* 3 (2011): 2–3.
104 Thomas Barnett, 'The Future of Fifth Generation Warfare: Follow the Food!' *Time*, June 8, 2011, available at: https://nation.time.com/2011/06/08/the-future-of-5th-generation-warfare-follow-the-food/
105 Marc E. Vargo, *The Weaponizing of Biology: Bioterrorism, Biocrime and Biohacking* (Jefferson, NC: McFarland, 2017), 45.
106 Derrick Tin, Pardis Sabeti, and Gregory R. Ciottone, 'Bioterrorism: An Analysis of Biological Agents Used in Terrorist Events,' *American Journal of Emergency Medicine* 54 (2022): 118.
107 David Malet, *Biotechnology and International Security* (Lanham, MD: Rowman & Littlefield, 2016), 107.
108 Ken Alibek, *Biohazard: The True Story of the Largest Covert Biological Weapons Programme in the World—Told from the Inside by the Man Who Ran It* (London: Random House, 1999), 259–260.
109 Alibek, *Biohazard*, 160.
110 Project for the New American Century, 'Rebuilding America's Defenses: Strategy, Forces and Resources for a New Century,' September 2000, available at: https://archive.org/details/RebuildingAmericasDefenses, 60.
111 National Academies of Sciences, *Biodefense in the Age of Synthetic Biology* (Washington, DC: National Academies of Sciences Press, 2018), 87.
112 Sarah Knapton, 'World Must Prepare for Biological Weapons That Target Ethnic Groups Based on Genetics, Says Cambridge University,' *The Telegraph*, August 12, 2019.
113 Malet, *Biotechnology and International Security*, 34.
114 Tom Mangold and Jeff Goldberg, *Plague Wars: The Terrifying Reality of Biological Warfare* (New York: St. Martin's Press, 1999), 244.
115 Wertheim, 'China's Push to Control Americans' Health Care Future.'
116 Paul Dabbar, 'U.S. Research Scientists Are Blind to China's Threat,' *Wall Street Journal*, April 4, 2023, A19.
117 U.S. Department of State, '2021 Adherence to and Compliance with Arms Control, Nonproliferation, and Disarmament Agreements and Commitments.'

118 Filippa Lentzos, 'How to Protect the World from Ultra-Targeted Biological Weapons,' *The Bulletin of Atomic Scientists*, December 7, 2020, available at: https://thebulletin.org/premium/2020-12/how-to-protect-the-world-from-ultra-targeted-biological-weapons/

119 Lentzos, 'How to Protect the World from Ultra-Targeted Biological Weapons.'

120 Malet, *Biotechnology and International Security*, 10.

121 DARPA is developing genetically modified mosquitoes in its 'Insect Allies' program, which are meant to genetically alter crops in the wild. This has raised concerns that the same technology could be used to destroy crops. Susan Scutti, 'Military Research Raises Concern about Bioterror Attack . . . by Insects,' *CNN*, October 5, 2018, available at: www.cnn.com/2018/10/05/health/insects-gene-editing-crops-darpa/index.html

122 Vladimir Belous, 'Weapons of the 21st Century,' *International Affairs* 55, no. 2 (2009): 80–81.

123 Makhmut Gareev, *If War Comes Tomorrow?* (Abingdon, UK: Routledge, 1995), 50.

124 S.G. Chekinov and S.A. Bogdanov, 'The Nature and Content of a New Generation War,' *Military Thought* 4 (2013): 14.

125 Timothy L. Thomas, 'Information Weapons: Russia's Nonnuclear Strategic Weapons of Choice,' *The Cyberdefense Review* (Summer 2020): 125–141.

126 Christopher Leake and Will Stewart, 'Putin Targets Foes with "Zombie" Gun Which Attacks Victims' Central Nervous System,' *Daily Mail*, March 31, 2012, available at: www.dailymail.co.uk/news/article-2123415/Putin-targets-foes-zombie-gun-attack-victims-central-nervous-system.html

127 Robert McCreight, 'Neuro-Cognitive Warfare: Inflicting Strategic Impact via Non-Kinetic Threat,' *Small Wars Journal*, September 16, 2022, available at: https://smallwarsjournal.com/jrnl/art/neuro-cognitive-warfare-inflicting-strategic-impact-non-kinetic-threat; original emphasis.

128 Ryan Clarke, Xiaxuo Sean Lin, and L.J. Eads, 'Enumerating, Targeting and Collapsing the Chinese Communist Party's NeuroStrike Program: Aggregating Intelligence Fragments and the Power of Network Graphs,' CCP Biothreat Initiative, June 2023, available at: https://static1.squarespace.com/static/6444894f2a886e74091c9e1b/t/6490791efa95ba0a3008ef1b/1687189791347/Enumerating%2C+Targeting+and+Collapsing+the+Chinese+Communist+Party%E2%80%99s+NeuroStrike+Program.pdf

129 Many recent riots and uprisings have been started by incidents of police brutality targeting a minority group, such as the George Floyd riots in 2020 and, more recently, the French riots of summer 2023.

130 Omar Shahabudin McDoom, 'Psychology of Threat in Intergroup Conflict: Emotions, Rationality, and Opportunity in the Rwandan Genocide,' *International Security* 37, no. 2 (2012): 132.

131 Mark Safranski, '5GW: Into the Heart of Darkness,' in: Abbott (ed), *The Handbook of 5GW*, 132.

132 McDoom, 'Psychology of Threat in Intergroup Conflict,' 142–146.

133 An interesting fact about the Rwandan genocide, recently discovered, is that supporters of the Rwandan regime imported 581 tons of machetes in the three years prior to the genocide. Roland Tissot, 'Did Machete Imports Prove that the Genocide against the Tutsi Was Planned?' *The Conversation*, September 2, 2020, available at: https://theconversation.com/did-machete-imports-to-rwanda-prove-that-the-genocide-against-the-tutsi-was-planned-145374

134 Andreas Maerker and Susanne Guski-Leinwand, 'Psychologists' Involvement in Repressive "Stasi" Secret Police Activities in Former East Germany,' *International Perspectives in Psychology* 7, no. 2 (2018): 109.

135 Maerker and Guski-Leinwand, 'Psychologists' Involvement in Repressive "Stasi" Secret Police Activities in Former East Germany,' 111.

136 Maerker and Guski-Leinwand, 'Psychologists' Involvement in Repressive "Stasi" Secret Police Activities in Former East Germany,' 112.
137 Sue Masterman, 'Report: Dissidents Tracked Using Radiation,' *ABC News*, January 4,
2001, available at: https://abcnews.go.com/International/story?id=81775&page=1
138 Alex Trembath and Vijaya Ramachandran, 'The Malthusians Are Back,' *The Atlantic*, March 22, 2023, available at: www.theatlantic.com/ideas/archive/2023/03/
population-control-movement-climate-malthusian-similarities/673450/
139 Klaus Schwab, *The Fourth Industrial Revolution* (New York: Currency, 2016),
92–93.

References

Albright, Katherine and McIntyre, Liz, *Spychips: How Major Corporations and Govern-
ment Plan to Track Your Every Move with RFID* (Nashville, TN: Nelson Current, 2005).
Alibek, Ken, *Biohazard: The True Story of the Largest Covert Biological Weapons Programme in the World—Told from the Inside by the Man Who Ran It* (London: Random
House, 1999).
Baard, Mark, 'Sentient World: War Games on the Grandest Scale,' *The Register*, June 23,
2007, available at: www.theregister.com/2007/06/23/sentient_worlds/
Bank for International Settlements, 'CBDCs: An Opportunity for the Monetary System,'
available at: www.bis.org/publ/arpdf/ar2021e3.pdf, accessed on April 22, 2022.
———, 'BIS Innovation Hub Work on Central Bank Digital Currency (CBDC),' available
at: www.bis.org/about/bisih/topics/cbdc.htm, accessed July 20, 2023.
———, 'III. Blueprint for the Future Monetary System: Improving the Old, Enabling the
New,' *BIS Annual Economic Report*, June 2023.
Banoun, Helene, 'mRNA: Vaccine or Gene Therapy? The Safety Regulatory Issues,' *International Journal of Molecular Sciences* 24, no. 13 (2023), https://doi.org/10.3390/
ijms241310514
Barnett, Thomas, 'The Future of Fifth Generation Warfare: Follow the Food!' *Time*, June 8,
2011, available at: https://nation.time.com/2011/06/08/the-future-of-5th-generation-warfare-
follow-the-food/
Belous, Vladimir, 'Weapons of the 21st Century,' *International Affairs* 55, no. 2 (2009): 64–82.
Benedikt, Carl and Osborne, Michael, 'The Future of Employment,' Oxford University Working Paper, September 17, 2013, available at: www.oxfordmartin.ox.ac.uk/downloads/
academic/future-of-employment.pdf
Beyer, Rebecca, 'Is Gene Editing Ethical? It Depends,' *New York University News*, February 26, 2019, available at: www.nyu.edu/about/news-publications/news/2019/february/
is-gene-editing-ethical—it-depends.html
Bland, Eric, 'Army Developing "Synthetic Telepathy,"' *NBC News*, October 13, 2008, avail-
able at: www.nbcnews.com/id/wbna27162401
Bostrom, Nick, 'Human Genetic Enhancements: A Transhumanist Perspective,' *The Journal
of Value Inquiry* 37 (December 2003): 493–506.
———, *Superintelligence: Paths, Dangers, Strategies* (Oxford: Oxford University Press,
2016).
Brenner, Katie, 'U.S. Charges Chinese Military Officers in 2017 Equifax Hacking,' *New
York Times*, February 10, 2020.
Cerri, Tony and Chaturvedi, Alok, 'A Sentient World Simulation (SWS): A Continuous
Running Model of the Real World/A Concept Paper for Comments,' Purdue University,

August 22, 2006, available at: https://business.purdue.edu/academics/MIS/workshop/papers/AC2_100606.pdf

Chaturvedi, Alok, 'Computational Challenges for a Sentient World Simulation,' Purdue University, March 10, 2006, available at: https://archive.ph/20060911223310/http:/www.purdue.edu/acsl/abstract/march10_06.html - selection-209.1–209.193

Chekinov, S.G. and Bogdanov, S.A., 'The Nature and Content of a New Generation War,' *Military Thought* 4 (2013): 12–23.

Chertoff, Michael, *Exploiting Data: Reclaiming Our Cyber Security in the Digital Age* (New York: Atlantic Monthly Press, 2018).

Clapper, James, 'Statement for the Record: Worldwide Threat Assessment of the US Intelligence Community,' Senate Armed Services Committee, February 9, 2016, available at: www.dni.gov/files/documents/SASC_Unclassified_2016_ATA_SFR_FINAL.pdf

Clarke, Ryan, Lin, Xiaxuo Sean, and Eads, L.J., 'Enumerating, Targeting and Collapsing the Chinese Communist Party's NeuroStrike Program: Aggregating Intelligence Fragments and the Power of Network Graphs,' CCP Biothreat Initiative, June 2023, available at: https://static1.squarespace.com/static/6444894f2a886e74091c9e1b/t/6490791efa95ba0a3008ef1b/1687189791347/Enumerating%2C+Targeting+and+Collapsing+the+Chinese+Communist+Party%E2%80%99s+NeuroStrike+Program.pdf

Clauser, Grant, 'Amazon's Alexa Never Stops Listening to You: Should You Worry?' *New York Times*, August 8, 2019.

Cole, Leonard A., *Clouds of Secrecy: The Army's Germ Warfare Tests over Populated Areas* (Totowa, NJ: Rowman & Littlefield, 1988).

Conger, Kate, 'Ex-Amazon Worker Convicted in Capital One Hacking,' *New York Times*, June 17, 2022.

Contera, Sonia, *Nano Comes to Life: How Nanotechnology Is Transforming Medicine and the Future of Biology* (Princeton, NJ: Princeton University Press, 2019).

Cooper, Anderson, 'Yuval Noah Harari on the Power of Data, Artificial Intelligence and the Future of the Human Race,' *CBS News*, October 31, 2021, available at: www.cbsnews.com/news/yuval-noah-harari-sapiens-60-minutes-2021-10-31/

Corbyn, Zoe, 'Facebook Experiment Boosts Voter Turnout,' *Nature*, September 12, 2012, available at: www.nature.com/articles/nature.2012.11401

Crespo, Ricardo A., 'Currency Warfare and Cyber Warfare: The Emerging Currency Battlefield of the 21st Century,' *Comparative Strategy* 37, no. 3 (2018): 236–237.

Cross, Glenn, 'Biological Weapons in the "Shadow War,"' *War on the Rocks*, November 9, 2021, available at: https://warontherocks.com/2021/11/biological-weapons-in-the-shadow-war/

Dabbar, Paul, 'U.S. Research Scientists Are Blind to China's Threat,' *Wall Street Journal*, April 4, 2023, A19.

Drexler, Eric, *Engines of Creation: The Coming Era of Nanotechnology* (New York: Anchor Books, 1987).

Edwards, Charlotte, 'Injectable Body Sensors Could Stream Medical Data to Smartphone,' *Medical Device Network*, March 19, 2018, available at: www.medicaldevice-network.com/news/injectable-body-sensors-stream-medical-data-smartphones/

Engelhardt, Tom and Greenwald, Glenn, *Shadow Government: Surveillance, Secret Wars, and a Global Security State in a Single Superpower World* (Chicago, IL: Haymarket Books, 2014).

Epstein, Robert, 'How Google Could Rig the 2016 Election,' *Politico*, August 19, 2015, available at: www.politico.com/magazine/story/2015/08/how-google-could-rig-the-2016-election-121548/

Farhani, Nita, Presentation at the World Economic Forum, January 20, 2023, available at: www.youtube.com/watch?v=VKg4NU3Pi2Y

Federal Reserve Bank of Boston, 'Project Hamilton Phase 1: A High Performance Payment Processing System Designed for Central Bank Digital Currencies,' February 3, 2022, available at: www.bostonfed.org/publications/one-time-pubs/project-hamilton-phase-1-executive-summary.aspx, accessed April 22, 2022.

Fildes, Nic, Bond, David, and Ram, Aliya, 'Cambridge Analytica under Fire as Scandal Grows,' *Financial Times*, March 20, 2018.

Gardner, M., 'The Internet of Bodies Will Change Everything, for Better or Worse,' *RAND*, October 29, 2020, available at: www.rand.org/blog/articles/2020/10/the-internet-of-bodies-will-change-everything-for-better-or-worse.html

Gareev, Makhmut, *If War Comes Tomorrow?* (Abingdon, UK: Routledge, 1995).

Halamka, John, Juels, Ari, Stubblefield, Adam, and Westhues, Jonathan, 'The Security Implications of VeriChip Cloning,' *Journal of the American Medical Informatics Association* 13, no. 6 (2006): 601–607.

Hammes, Thomas X., 'Fourth Generation Warfare Evolves, Fifth Emerges,' *Military Review* 87, no. 3 (2007): 14–23.

Helbing, Dirk and Sánchez-Vaquerizo, Javier Argota, 'Digital Twins: Potentials, Ethical Issues, and Limitations' in: Andrej Zwitter and Oskar Gstrein, *Handbook on the Politics and Governance of Big Data and Artificial Intelligence* (Cheltenham, UK: Edward Elgar, 2023), available at: https://papers.ssrn.com/sol3/papers.cfm?abstract_id=4167963

Holder, Josh, 'Tracking Coronavirus Vaccinations around the World,' *New York Times*, March 13, 2023, available at: www.nytimes.com/interactive/2021/world/covid-vaccinations-tracker.html

Hugh-Jones, Martin E., Hatch Rosenberg, Barbara, and Jacobsen, Stuart, 'The 2001 Anthrax Attack: Key Observations,' *Bioterrorism & Biodefense* 3 (2011), doi:10.4172/2157–2526. S3–008

Huxley, Aldous, *Brave New World Revisited* (New York: Harper Perennial, 2006).

Huxley, Julian, 'Transhumanism,' *Journal of Humanistic Psychology* 8, no. 1 (1968): 73–76.

Joy, Bill, 'Why the Future Does Not Need Us,' *Wired Magazine* 8.04 (2000), available at: www.cc.gatech.edu/computing/nano/documents/Joy - Why the Future Doesn't Need Us.pdf

Jung, Hyunyun, 'Blockchain Implementation Method for the Interoperability between CBDCs,' *Future Internet* 13, no. 5 (2021): 1–14.

Key, Alys, 'Nigeria Limits Cash Withdrawals to $45 per Day in CBDC, Digital Banking Push,' Yahoo Finance, December 7, 2022, available at: https://finance.yahoo.com/news/nigeria-limits-cash-withdrawals-45-100611213.html

Khurana, Amit, Prince Allawahdi, Khurana, Isha, Allwahdi, Sachin, Weiskirchen, Ralf, Banothu, Anil Kumar, Chhabra, Deepak, Joshi, Kamaldeep, and Bharani, Kala Kumar, 'Role of Nanotechnology behind the Success of mRNA Vaccines for COVID-19,' *Nano Today* 38 (2021): 1–6.

Knapton, Sarah, 'World Must Prepare for Biological Weapons That Target Ethnic Groups Based on Genetics, Says Cambridge University,' *The Telegraph*, August 12, 2019.

Koblentz, Gregory, *Living Weapons: Biological Warfare and International Security* (Ithaca, NY: Cornell University, 2009).

Koehler, John O., *Stasi: The Untold Story of the East German Secret Police* (Boulder, CO: Westview Press, 1999).

Kouzminov, Alexander, *Biological Espionage: Special Operations of the Soviet and Russian Intelligence Services in the West* (London: Greenhill Books, 2005).

Kurzweil, Ray, *The Age of Spiritual Machines* (New York: Penguin, 2000).

Leake, Christopher and Stewart, Will, 'Putin Targets Foes with "Zombie" Gun Which Attacks Victims' Central Nervous System,' *Daily Mail*, March 31, 2012, available at: www.dailymail.co.uk/news/article-2123415/Putin-targets-foes-zombie-gun-attack-victims-central-nervous-system.html

Lee, Kai-Fu, *AI Superpowers: China, Silicon Valley, and the New World Order* (Boston: Houghton Mifflin Harcourt, 2018).

Lentzos, Filippa, 'How to Protect the World from Ultra-Targeted Biological Weapons,' *The Bulletin of Atomic Scientists*, December 7, 2020, available at: https://thebulletin.org/premium/2020-12/how-to-protect-the-world-from-ultra-targeted-biological-weapons/

LeVaux, Ari, 'Genetically Modified Food May Pose a Real Danger to Health,' in: Noel Merino (ed.), *Genetically Modified Food* (Farmington Hills, MI: Greenhaven Press, 2014), 111–118.

Liao, S. Matthew, Sandberg, Anders, and Roache, Rebecca, 'Human Engineering and Climate Change,' *Ethics, Policy, and Environment* 15, no. 2 (2012): 208–211.

Lichter, Andreas, Löffler, Max, and Siegloch, Sebastian, 'The Long-Term Costs of Government Surveillance: Insights from Stasi Spying in East Germany,' *Journal of the European Economic Association* 19, no. 2 (2021): 741–789.

Lunev, Stanislav, *Through the Eyes of the Enemy: Russia's Highest Ranking Military Defector Reveals Why Russia Is More Dangerous Than Ever* (Washington, DC: Regnery, 1998).

Maerker, Andreas and Guski-Leinwand, Susanne, 'Psychologists' Involvement in Repressive "Stasi" Secret Police Activities in Former East Germany,' *International Perspectives in Psychology* 7, no. 2 (2018): 107–119.

Malet, David, *Biotechnology and International Security* (Lanham, MD: Rowman & Littlefield, 2016).

Mangold, Tom and Goldberg, Jeff, *Plague Wars: The Terrifying Reality of Biological Warfare* (New York: St. Martin's Press, 1999).

Masterman, Sue, 'Report: Dissidents Tracked Using Radiation,' *ABC News*, January 4, 2001, available at: https://abcnews.go.com/International/story?id=81775&page=1

Masunaga, Samantha, 'A Quick Guide to Elon Musk's New Brain-Implant Company, Neuralink,' *Los Angeles Times*, April 21, 2017.

McCreight, Robert, 'Neuro-Cognitive Warfare: Inflicting Strategic Impact via Non-Kinetic Threat,' *Small Wars Journal*, September 16, 2022, available at: https://smallwarsjournal.com/jrnl/art/neuro-cognitive-warfare-inflicting-strategic-impact-non-kinetic-threat

McDoom, Omar Shahabudin, 'Psychology of Threat in Intergroup Conflict: Emotions, Rationality, and Opportunity in the Rwandan Genocide,' *International Security* 37, no. 2 (2012): 119–155.

Moran, Andrew, 'CBDCs with Expiration Dates, Restrictions Could Target Social Policies, Economist Tells WEF,' *Epoch Times*, June 30, 2023, available at: www.theepochtimes.com/expiry-dates-restrictions-benefits-of-central-bank-digital-currencies-economist-tells-wef_5366056.html

Moravec, Hans, *Robot: Mere Machine to Transcendent Mind* (Oxford: Oxford University Press, 1999).

Nakashima, Ellen, 'Hacks of OPM Databases Compromised 22.1 Million People, Federal Authorities Say,' *Washington Post*, July 9, 2015.

National Academies of Sciences, *Biodefense in the Age of Synthetic Biology* (Washington, DC: National Academies of Sciences Press, 2018).

National Security Commission on Artificial Intelligence, Final Report, March 2021, available at: www.nscai.gov/wp-content/uploads/2021/03/Full-Report-Digital-1.pdf

Newman, Stuart A., 'The Transhumanism Bubble,' *Capitalism Nature Socialism* 21, no. 2 (2010): 29–42.

New York Times, 'Obama to Unveil Initiative to Map the Human Brain,' *New York Times*, April 2, 2013.

Nicas, Jack, Isaac, Mike, and Swanson, Ana, 'U.S. Opens Review into China's TikTok,' *New York Times*, November 2, 2019.

Perlroth, Nicole, 'All 3 Billion Yahoo Accounts Were Affected by 2013 Attack,' *New York Times*, October 3, 2017.

Pierce, David, 'The All-Seeing Kinect: Tracking My Face, Arms, Body, and Heart on the Xbox One,' *The Verge*, May 21, 2013, available at: www.theverge.com/2013/5/21/4353232/kinect-xbox-one-hands-on

Piore, Adam, 'The Neuroscientist Who Wants to Upload Humanity to a Computer,' *Popular Science*, May 16, 2014, available at: www.popsci.com/article/science/neuroscientist-who-wants-upload-humanity-computer/

Project for the New American Century, 'Rebuilding America's Defenses: Strategy, Forces and Resources for a New Century,' September 2000, available at: https://archive.org/details/RebuildingAmericasDefenses

Rana, Fazale R. with Samples, Kenneth R., *Humans 2.0: Scientific, Philosophical, and Theological Perspectives on Transhumanism* (Covina, CA: RTB Press, 2019).

Rao, Rajesh P.N., Stocco, Andrea, Bryan, Matthew, Sarma, Devaparatim, Youngquist, Tiffany M., Wu, Joseph, and Prat, Chantal S., 'A Direct Brain-to-Brain Interface in Humans,' *PLOS One* 9, no. 11 (2014): 1–12.

Rawnaque, Ferdousi Sabera, Rahman, Khandoker Mahmudur, Anwar, Syed Ferhat, Vaidyanathan, Ravi, Chau, Tom, Sarker, Farhana, and Al Mamun, Khondaker Abdullah, 'Technological Advancements and Opportunities in Neuromarketing: A Systematic Review,' *Brain Inform* 7, no. 1 (2020): 1–19.

Regalado, Antonio, 'The Creator of the CRISPR Babies Has Been Released from a Chinese Prison,' *MIT Technology Review*, April 4, 2022, available at: www.technologyreview.com/2022/04/04/1048829/he-jiankui-prison-free-crispr-babies/

Rosoff, Matt, 'A.I. May Replace Most of Today's Jobs, but This Start-Up Investor Sees Global GDP Increasing by 50 Percent a Year within Decades,' *CNBC*, February 26, 2019, available at: www.cnbc.com/2019/02/26/sam-altman-on-ai-jobs-may-go-away-but-massive-abundance-likely.html

Scharre, Paul, *Four Battlegrounds: Power in the Age of Artificial Intelligence* (New York: W.W. Norton, 2023).

Schneier, Bruce, *Click Here to Kill Everybody: Security and Survival in a Hyper-connected World* (New York: W.W. Norton, 2018).

Schwab, Klaus, *The Fourth Industrial Revolution* (New York: Currency, 2016).

Scutti, Susan, 'Military Research Raises Concern about Bioterror Attack . . . by Insects,' CNN, October 5, 2018, available at: www.cnn.com/2018/10/05/health/insects-gene-editing-crops-darpa/index.html

Sly, Liz, 'U.S. Soldiers Are Revealing Sensitive and Dangerous Information by Jogging,' *Washington Post*, January 28, 2018.

Solimano, Pedro, 'Brazilian CBDC Allows Government to Freeze Funds, Developer Finds,' Yahoo Finance, July 16, 2023, available at: https://finance.yahoo.com/news/brazilian-cbdc-allows-government-freeze-174630324.html

Tangermann, Victor, 'Elon Musk Says Neuralink's First Product Will Control Smartphone with Brain Implant,' *Futurism*, April 9, 2021, available at: https://futurism.com/elon-musk-neuralink-first-product-control-smartphone-brain-implant

Thomas, Timothy L., 'Information Weapons: Russia's Nonnuclear Strategic Weapons of Choice,' *The Cyberdefense Review* (Summer 2020): 125–141.

Tin, Derrick, Sabeti, Pardis, and Ciottone, Gregory R., 'Bioterrorism: An Analysis of Biological Agents Used in Terrorist Events,' *American Journal of Emergency Medicine* 54 (2022): 117–121.

Tissot, Roland, 'Did Machete Imports Prove that the Genocide against the Tutsi Was Planned?' *The Conversation*, September 2, 2020, available at: https://theconversation.com/did-machete-imports-to-rwanda-prove-that-the-genocide-against-the-tutsi-was-planned-145374

Trembath, Alex and Ramachandran, Vijaya, 'The Malthusians Are Back,' *The Atlantic*, March 22, 2023, available at: www.theatlantic.com/ideas/archive/2023/03/population-control-movement-climate-malthusian-similarities/673450/

University of Leeds, C40, and Arup, 'The Future of Urban Consumption in a 1.5°C World: C40 Cities Headline Report,' June 2019, available at: www.c40.org/wp-content/uploads/2021/08/2270_C40_CBE_MainReport_250719.original.pdf

U.S. Department of State, '2021 Adherence to and Compliance with Arms Control, Nonproliferation, and Disarmament Agreements and Commitments,' April 2021, available at: www.state.gov/wp-content/uploads/2021/04/2021-Adherence-to-and-Compliance-With-Arms-Control-Nonproliferation-and-Disarmament-Agreements-and-Commitments.pdf

Vargo, Marc E., *The Weaponizing of Biology: Bioterrorism, Biocrime and Biohacking* (Jefferson, NC: McFarland, 2017).

Vigna, Paul and Casey, Michael J., *The Age of Cryptocurrency: How Bitcoin and Digital Money Are Challenging the Global Economic Order* (New York: St. Martin's Press, 2015).

Wakefield, Jane, 'People Devote Third of Waking Time to Mobile Apps,' *BBC News*, January 12, 2022, available at: www.bbc.com/news/technology-59952557

Weiss, Haley, 'Why You Are Probably Getting a Microchip Implant Someday,' *The Atlantic*, September 21, 2018, available at: www.theatlantic.com/technology/archive/2018/09/how-i-learned-to-stop-worrying-and-love-the-microchip/570946/

Werner, Richard, 'Crucial CBDCs with Professor Richard Werner: YOUR Future Is Being Decided,' available at: www.youtube.com/watch?v=KGpQLbZXKME, accessed on July 20, 2023.

Wertheim, Jon, 'China's Push to Control Americans' Health Care Future,' *CBS News*, January 31, 2021, available at: www.cbsnews.com/news/biodata-dna-china-collection-60-minutes-2021-01-31/

Whitehouse-Levine, Miller, 'CBDCs Will Lead to Absolute Government Control,' *Cointelegraph*, April 6, 2023, available at: https://cointelegraph.com/news/desantis-is-right-cbdcs-will-lead-to-absolute-government-control

Wood, Patrick, *The Evil Twins of Technocracy and Transhumanism* (Meza, AZ: Coherent, 2022).

World Health Organization, 'The European Commission and WHO Launch Landmark Digital Health Initiative to Strengthen Global Health Security,' Press Release, June 5, 2023, available at: www.who.int/news/item/05-06-2023-the-european-commission-and-who-launch-landmark-digital-health-initiative-to-strengthen-global-health-security

Wylie, Christopher, *Mindf*ck: Cambridge Analytica and the Plot to Break America* (New York: Random House, 2019).

Zuboff, Shoshana, *The Age of Surveillance Capitalism: The Fight for the Future at the New Frontier of Power* (London: Profile Books, 2019).

6 Towards the Sixth Generation of Warfare

As the current fifth generation war over global domination is waged in the informational and cognitive domain by supra-combinations of state and nonstate actors, with an outcome that is still uncertain, a new generation warfare is on the horizon. Russian military theorists have discussed concepts such as new generation warfare, non-contact warfare, nonlinear warfare, and sixth generation warfare since the 1990s. These concepts overlap somewhat with ideas presented in this chapter but are based on a different understanding of the history and theory of war.[1] The sixth generation of war in Russian theory would be characterized by non-contact operations that rely on high-precision weapons, information weapons, weapons based on new physical principles, and electronic warfare.[2]

Of particular interest here are the somewhat mysterious new physical principle (NPP) weapons, as they are secret weapons that are meant to achieve an asymmetric advantage over an adversary and deny an adversary the ability to benefit from a conventional or nuclear military advantage. According to Stephen Blank, NPP weapons would include 'beam, geophysical, genetic, psychophysical and other technology,' and 'they will [in Putin's view] become the main means for achieving a decisive victory, including in global conflicts.'[3] Peter Vincent Pry has pointed out that some U.S. analysts have called the radical asymmetric strategies pursued by Russia, China, and Iran 'Cybergeddon,' 'Blackout War,' or 'Combined-Arms Cyber Warfare.'[4] This chapter will refer to these ideas simply as 6GW.

Contours of 6GW

At this time, there is hardly any literature on what 6GW would look like; there are only different ideas as to how one could conceptualize the next evolution of war. Patrick Dugan has argued that '6GW involves the systematic knowledge tapped by 5GW, but leveraged by the capability of a mind to self-modify. If 5GW is contextual warfare then 6GW is design warfare, a contest of systemic resilience.'[5] According to a Pakistani military blog, 6GW

> seeks to disrupt the enemy's ability to mobilize forces, identify and target goals and objectives, and deny access to critical resources. As such it works

DOI: 10.4324/9781003396963-7

to counter and prevent traditional means of warfare, such as conventional military forces and tactics.[6]

What will be presented here is speculative and mostly represents the author's interpretations of Russian and Chinese ideas of asymmetric attack rather than any consensus view of what 6GW is in the framework of generational warfare.

Sixth Generation Warfare in the xGW Framework

According to the logic of the xGW theory, each higher gradient of warfare would be more complex, and violence would be more dispersed or more indirect than in lower gradients. 'In the xGW framework, kinetic force has greater utility at the lower gradients of the framework and less utility at the higher gradients of the framework.'[7] Each subsequent gradient would also be a response to overcome the previous gradient.[8] Since 4GW's paradigm was insurgency, the paradigm of 5GW is, hence, counterinsurgency. It follows that 6GW must overcome or neutralize the 5GW paradigm. Sixth generation warfare shifts warfare from the human domain to the domain of technical systems. Just as 4GW and 5GW have bypassed traditional military power by focusing on the human domain in conflict, 6GW bypasses the human domain by attacking technical systems that are the foundations that underlie modern civilizations, as well as military power, and that have become critical to human survival.

Similar to 5GW, 6GW will be fought mostly in secret, but the effects will have greater visibility. Unlike 5GW, the capabilities necessary for 6GW cannot be found in the private sector but come from the highly secretive world of military research programs and intelligence agencies with a technical focus. Sixth generation warfare is enabled by DARPA and its counterparts in China, Russia, and other technologically advanced states.[9] Furthermore, although 6GW seeks to make the use of military force impossible by targeting technical systems, it has the potential to be substantially more violent than 5GW. However, the violence will be likely disguised as glitches, accidents, or natural disasters. People will be harmed primarily through secondary or tertiary effects of attacks rather than by attacks that directly target humans. The most likely cause of death in 6GW will be starvation, as technical systems that are essential to the production and distribution of food will fail and as few people in the world are self-sufficient.

The 6GW Paradigm Is Strategic Sabotage

Sabotage is defined as 'the act of destroying or damaging something deliberately so that it does not work correctly.'[10] It is a practice commonly used in wartime in an effort to undermine an adversary's ability to use all available resources and, hence, to indirectly degrade the enemy's ability to employ force. Sabotage can target economic capabilities (factories, warehouses), infrastructure (railways, roads, bridges, pipelines), military capabilities (radar, military equipment, munition depots), food production (agriculture, livestock), and organizations (political bodies, labor unions). Sabotage is particularly a Special Operations Forces mission (usually included in direct action) that is defined in the following way:

> an unconventional warfare activity involving an act(s) intended to injure, interfere with, or obstruct the national defense of a country by willfully injuring or destroying, or attempting to injure or destroy, any national defenses or war material, premises, or utilities, including human and natural resources. Sabotage selectively disrupts, destroys, or neutralizes hostile capabilities with a minimum expenditure of manpower and material.[11]

The benefits of sabotage during peacetime are much more limited because the effects tend to be temporary since sabotaged infrastructure or machinery can be fixed or replaced.[12] Furthermore, if sabotage was discovered and attributed, it would amount to an act of war and could result in the victimized state seeking to retaliate or escalating a covert conflict into a full-scale war.[13] Given the limited benefits and high stakes of peacetime sabotage actions, it is often assumed that sabotage would only occur in wartime. However, sabotage may be seen as beneficial in the 'gray zone' between war and peace, which is 'characterized by ambiguity about the nature of the conflict, opacity of the parties involved, or uncertainty about the relevant policy and legal frameworks.'[14]

As revealed in the *Mitrokhin Archive*, the Soviet KGB made extensive plans for the systematic sabotage of 'power-transmission lines, oil pipelines, communications systems and major industrial complexes in most, if not all, NATO countries.'[15] According to Russian defector Stanislav Lunev, Soviet Spetsnaz (Special Forces) would, during wartime, 'try to assassinate as many American leaders as possible, as well as their families. They would also blow up power stations, telephone switching systems, dams, and any strategic target that cannot be taken out with long range weapons.'[16] He suggested that Russian Special Forces regularly entered the United States with fake passports, posing as tourists from Eastern Europe. Lunev also supported the claim made by General Alexander Lebed that KGB suitcase nukes had remained unaccounted for, alleging that they may have been pre-deployed in the West to be used in the event of war.[17] Although these plans were obviously never carried out, it is quite possible that China and Russia may have already prepared for strategic sabotage in the West in anticipation of war, and that they are already conducting a covert sabotage campaign designed to weaken the United States and certain other Western countries.[18] In today's world, there are several approaches for conducting strategic sabotage:

- Encourage sabotage by groups in a target society that can be co-opted.
- Insert professional saboteurs into the target society (intelligence or Special Forces personnel).
- Use cyber warfare to covertly sabotage critical infrastructure.
- Embed exploitable vulnerabilities in technical equipment sold to adversaries.
- Modify the environment to disrupt a target society.

Objectives in 6GW

The primary objective in 5GW is to dominate the human domain by covertly manipulating proxies and influencing individuals to unwittingly act on behalf of a 5GW opponent or to make decisions that benefit the 5GW opponent. In 6GW, the primary objective will be to sabotage or subvert technical systems that are critical to human survival. Just like in 5GW, control of escalation and avoidance of direct military conflict will be of great concern to 6GW belligerents. They have to find or create vulnerabilities in the enemy's technical infrastructure that they can exploit in a manner that it does not look like an act of aggression or in a manner that makes correct attribution very unlikely. The main target for 6GW attacks is key infrastructure related to telecommunications, energy, the electric grid, transportation, and food production. Should one of these critical infrastructure systems fail in a target society, it would not only cause severe disruption but would also endanger the survival of the targeted society.

Telecommunications Infrastructure

Telecommunications are critical to the functioning of modern society, which has become extremely dependent on the internet.[19] As of 2021, there were 4.2 billion people in the world with internet access, which represents more than half of the world's population.[20] Shopping, banking, media, and personal communications have all largely been moved online. An internet outage would severely affect a country's economy as online shopping would become unavailable and as ATMs would no longer work.[21] It would affect the financial system as credit card payments, bank transfers, and trading on stock and commodity markets would be disrupted. It would impact logistics systems and travel booking systems, disrupting the movement of goods and people's ability to travel. Since people in countries with high internet penetration receive most of their news online, the spread of information in a society would be disrupted, creating uncertainty and perhaps even panic. Although a worldwide internet outage would be very unlikely, there is the potential for nationwide or regional outages due to large-scale DDOS attacks, physical sabotage of underwater or land internet cables, electromagnetic jamming of wireless/satellite networks, and internet kill switches.

Energy Infrastructure

No country can manage without access to energy sources such as natural oil and gas, which are needed for transportation, heating/cooling, industrial production, and even agriculture. In the United States, 80 percent of all energy consumed is derived from fossil fuels, with the remainder coming from renewables and nuclear energy.[22] Few countries in the world are energy self-sufficient. Cutting off adversaries' access to cheap energy or diminishing their ability to refine, transport, and utilize energy can threaten their ability to function as industrial production and transportation could come to a standstill. Lack and/or high cost of energy can result in serious loss of life in winter in colder climate zones. Key energy infrastructure that could be targeted for sabotage includes refineries, pipelines, oil tankers, LNG and oil terminals, and oil and gas storage facilities. Objectives can be to deprive targeted countries of access to cheap energy, to disrupt the ability of an energy exporter to profit from energy exports, or to cause an environmental disaster that inflicts long-term environmental and economic damage.[23]

Electric Grid

All modern societies are critically dependent on a functioning electric grid for transportation (traffic signals, mass transportation systems, pumps at gas stations, and electric vehicles), industrial production, heating and cooling, lighting, the internet, telephone and mobile communications, access to finance (credit cards and ATMs), running water, and much more. An extended power outage can lead to increased crime in the form of vandalism and looting.[24] Within hours of a widespread power outage, tap water would stop, and people could die from dehydration after three days (if there was no rain or nearby freshwater sources). The EMP Commission has pointed out that the electric grid is vulnerable to four different modes of attack—sabotage of transformers and power lines, cyberattacks that seek to overload large power transformers in order to damage them, EMP-type attacks that destroy electronics and interfere with electrical devices over a large area, and weaponization of the weather—all of which can be termed 'blackout warfare.'[25]

Transportation Infrastructure

Adversaries can target the following elements of the transportation infrastructure with physical or cyber sabotage—trucking, shipping, the railway system, and air transportation. Sabotage can target ports, railroad tracks, train signals, computerized control systems for trains, air traffic control systems, computerized logistics and booking systems, and key components or parts needed for transportation such as truck tires and diesel engine oil.[26] The objective might be to disrupt supply chains, disrupt other economic activity, cause environmental damage, or cause distress to a population. Transportation systems are vulnerable to both physical and cyber sabotage, and their disruption can affect a society in numerous ways, including the production and distribution of goods, most importantly food, as well as key consumer goods and parts essential for running the economy.

Food Infrastructure

Food is absolutely critical to sustaining human life, and food shortages are known to cause political instability through increased protest, civil unrest, and general lawlessness.[27] This means that starvation is a tool of warfare that can be highly effective in terms of internally destabilizing the society of an adversary.[28] The more a nation is dependent on food imports, the more it is vulnerable to global supply

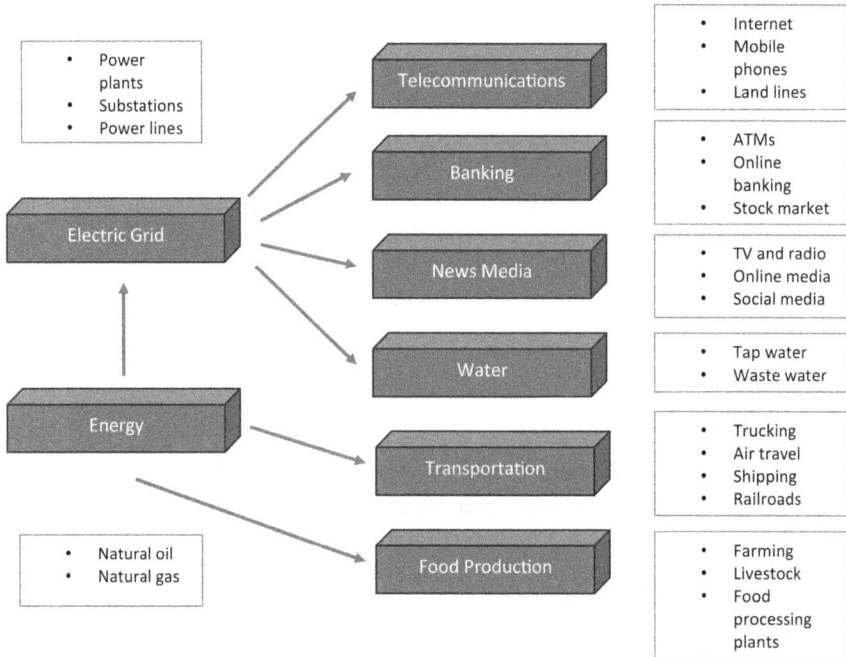

Figure 6.1 Critical Infrastructure
Source: Created by author

chain disruptions in the food sector.[29] Even nations that are able to produce most of their food are vulnerable to sabotage attacks that target their agricultural production, as there are limitations to the amount of food that can be stockpiled for dealing with shortages and also to the duration food can be stockpiled without spoiling. The United States had a strategic grain reserve in the 1970s but, currently, it no longer has any significant stockpiles of food.[30] For the most part, populations are dependent on fresh food that is consumed shortly after it has been produced. An adversary has a number of options as to how to interfere with the food supply and to create food shortages by way of sabotage. There can be biological attacks on crops and livestock, sabotage attacks directed at food plants and fertilizer plants, and attacks on crops involving setting fields of crops on fire or damaging them with chemical spills.

Cyber Warfare

Cyberspace has become an important domain of warfare and covert operations. The cyber domain enables remote attacks at lightning speed that are very difficult to attribute and might not be distinguishable from benign inputs.[31] Nation states seeking to conduct strategic sabotage through cyberspace may rely on nonstate cyber proxies such as patriotic hacker groups, criminal hackers, or cyber militias in order to make attribution more difficult. For example, as Thomas Rid pointed out, although it is commonly assumed that the Russian government masterminded the cyberattacks on Estonia in 2007, '[n]either experts from the Atlantic Alliance nor from the European Commission were able to identify Russian fingerprints in the operations.'[32] According to Kim Zetter, '[u]nless a hacker is sloppy about hiding his tracks, it's often not possible to unmask the perpetrator through digital evidence alone.'[33] CIA cyber tools leaked by WikiLeaks under the codename 'Vault 7' indicated that leading cyber powers such as the United States may 'steal' attack techniques from adversaries in order to misdirect forensic analysis.[34]

Critical Infrastructure Cyberattacks

The possibility of a cyberattack damaging machinery was first proven in an experiment conducted by the Department of Energy in 2007, which is known as the Aurora Generator Test. Just 21 lines of malicious code sent to a 2.25-megawatt generator caused it to spin out of control and break down.[35] The real-world test came with the Stuxnet attack, which was discovered in June 2010. Stuxnet targeted the Iranian SCADA (supervisory control and data acquisition) system that controlled centrifuges used for uranium enrichment, with the objective of causing moderate damage over time, which the Iranians would attribute to mechanical issues rather than sabotage, in order to slow down Iran's nuclear weapons program. The Stuxnet worm was so sophisticated that it masked the damage by providing feedback that the centrifuges were spinning normally.[36] The malware was only discovered by a computer security company in Belarus when it unintentionally spread around the world, likely owing to somebody connecting an infected USB drive to a computer connected to the internet.[37] Since Stuxnet, there have been numerous cases of cyberattacks that have targeted critical infrastructure, some of which were successful.

Energy Infrastructure Target

In 2012, Iranian hackers disrupted the business operations of Saudi Aramco, the biggest oil company in the world, for two weeks by planting Shamoon malware on its business network; the malware wiped the boot sector of 30,000 Saudi Aramco computers and made them unbootable.[38] In early May 2021, Colonial Pipeline was hit with a ransomware attack that forced the company to shut down a 5,500-mile pipeline that transports gasoline, diesel, and jet fuel from the Gulf Coast to New York.[39] The hack disrupted the availability of gasoline on the East Coast for several weeks. The Biden administration has claimed that the hackers responsible were based in Russia, and that the Russian intelligence service FSB has a history of co-opting and cooperating with criminal hackers, which may indicate a political rather than monetary motive for the hack.[40]

Transportation Infrastructure Target

The Danish company Maersk suffered a major cyberattack that shut down a terminal at the Port of Los Angeles in 2017, costing the shipping company between $200 million and $300 million owing to the disruption in its operations. The terminal was shut down for about five days. The problem was caused by Russian ransomware called NotPetya, which exploited a vulnerability in Microsoft Windows.[41] More recently, the Norwegian shipping company DNV was hit with a ransomware attack that affected 1,000 vessels, which represents 15 percent of its fleet. The weak spot was, again, logistics software that keeps track of cargo.[42] Future cyberattacks against shipping could target the computer systems of large vessels directly and could enable hackers to gain control of a ship's functions. Railway transportation is also vulnerable. In 2017, German Deutsche Bahn computers were infected with malware, causing incorrect displays at train stations and disrupting train traffic for several hours. A cyberattack on a train logistics company brought trains in Denmark to a standstill in November 2022. In October 2022, Russian hackers shut down airline websites in the United States with DDOS attacks targeting different U.S. airports, resulting in longer airport wait times and congestion.[43] On January 11, 2023, Federal Aviation Administration computer networks suffered from a supposed glitch that caused thousands of flights to be delayed and numerous flight cancellations.[44]

Food Infrastructure Target

Even the food supply is increasingly vulnerable to cyberattacks. In April 2022, the FBI issued a warning that there were 'ransomware attacks during these [planting and harvest] seasons against six grain cooperatives during the fall 2021 harvest and two attacks in early 2022 that could impact the planting season by disrupting the supply of seeds and fertilizer.' Furthermore, the FBI suggested that

> [a] significant disruption of grain production could impact the entire food chain, since grain is not only consumed by humans but also used for animal feed. In addition, a significant disruption of grain and corn production could impact commodities trading and stocks. An attack that disrupts processing at a protein or dairy facility can quickly result in spoiled products and have cascading effects down to the farm level as animals cannot be processed.[45]

Several major cyberattacks against food processing plants have been reported over the last few years. An attack against Dole Food Company forced the company to temporarily shut down production, which caused shortages of salads in Texas. A CNN article on the attack also pointed out that '[o]ther high-profile hacks against the food and agriculture sector in the last two years have threatened supply chains and caused distributors to strengthen their cybersecurity,' which included a ransomware attack by Russian hackers against JBS, the world's largest meat supplier, that forced the company to temporarily suspend production.[46]

Cyberattacks Targeting the Power Grid

The vulnerability of the power grid to cyberattacks is now widely understood and was warned about by Richard Clarke and Robert Knake as early as 2010.[47] Hackers could gain access to computerized control systems that manage the electric grid. A 2012 National Academies of Sciences study suggested that

> all such [SCADA] systems are potentially vulnerable to cyber attacks, whether through Internet connections or by direct penetration at remote sites. Any telecommunication link that is even partially outside the control of the system operators is a potentially insecure pathway into operations and a threat to the grid. If they could gain access, hackers could manipulate SCADA systems to disrupt the flow of electricity, transmit erroneous signals to operators, block the flow of vital information, or disable protective systems.[48]

That this possibility was not merely hypothetical was demonstrated in late 2014 by Russian hackers who managed to disrupt the power grid in Ukraine and left 225,000 Ukrainians without electricity for several hours in the middle of winter.[49]

Large Power Transformers

Ted Koppel has argued that large power transformers (LPTs) could be irreparably damaged in a cyberattack by being overloaded with electricity. LPTs are custom-built, are very bulky and difficult to transport, and cost $3 million–$10 million, and power companies do not have spares in case a LPT is damaged. Furthermore, 75 percent of them are imported from other countries, and the average delivery time is two years.[50] Particularly concerning is that, between 2009 and 2019, the United States imported over 200 LPTs from China; they are part of the U.S. electric grid and could have vulnerabilities the Chinese could exploit for large-scale sabotage.[51]

U.S. Cyberattacks on Electric Grid

The U.S. government developed a plan called Nitro Zeus, alongside Stuxnet, which concerned large-scale cyber sabotage of critical Iranian infrastructure, including Iran's telecommunications infrastructure and its electric grid. According to David Sanger and Mark Mazzetti, Nitro Zeus was an 'effort to infuse Iran's computer networks with "implants" that could be used to monitor the country's activities and, if ordered by Mr. Obama, to attack its infrastructure.'[52] In 2019, Venezuela suffered unprecedented, widespread power outages that President Maduro claimed were caused by U.S. cyberattacks, which was instantly denied by the U.S. government. An article in *Forbes* argued that

> the idea of a government like the United States remotely interfering with its power grid is actually quite realistic. Remote cyber operations rarely require a significant ground presence, making them the ideal deniable influence operation. Given the U.S. government's longstanding concern with Venezuela's

government, it is likely that the U.S. already maintains a deep presence within the country's national infrastructure grid, making it relatively straightforward to interfere with grid operations.[53]

Russian Malware in the U.S. Electric Grid

The *New York Times* reported, in 2019, that Russia had inserted malware in critical U.S. infrastructure to sabotage it in the event of war, and that the United States had also inserted malware to probe and attack the Russian electric grid since 2012.[54] Russian hackers keep probing the U.S. electric grid, which means that the Russian government could still decide to attack it in the case of a further escalation in Ukraine. Although a major cyberattack on the U.S. electric grid has yet to occur, it is increasingly threatened by physical attacks, perpetrated mostly by unknown parties and for unknown reasons.[55]

EMP Weapons

An electromagnetic pulse is one of the effects caused by the detonation of nuclear weapons. This effect was expected and observed during the first U.S. nuclear test at the Trinity site in July 1945, but was insufficiently understood in terms of its destructive potential. Only in 1962 did the high-altitude Starfish nuclear blast reveal the extraordinary range of the EMP as it knocked out electricity in Hawaii.[56] The EMP caused by this nuclear test was powerful enough to destroy electronic components hundreds of miles away.[57] According to Peter Vincent Pry, high-altitude EMPs (HEMPs) could be an attractive attack option for U.S. adversaries, as a 'HEMP attack detonates in outer space, leaving no collectable bomb debris. No fingerprints. HEMP attack might be executed anonymously, to escape retaliation.'[58] How vulnerable is the United States to an EMP attack?

The EMP Threat

The EMP threat has been discussed and warned of for decades. The Congressional EMP Commission was active from 2001 to 2008 and from 2017 to 2018. It produced several reports that clearly indicated how vulnerable American society is to natural or deliberate disruptions of the U.S. power grid caused by severe weather events, solar storms (Carrington event), EMP, or cyberattacks. The EMP Commission argued that even temporary, partial disruptions of the electric grid could cause severe damage, stating that '[i]t is possible for the functional outages to become mutually reinforcing until at some point the degradation of the infrastructure could have irreversible effects on the country's ability to support the population.'[59] In its 2017 report, it gave a very severe warning, suggesting that U.S. adversaries would have both the capability and the intent to attack the U.S. electric grid:

> Combined-arms cyber warfare, as described in the military doctrines of Russia, China, North Korea, and Iran, may use combinations of cyber-, sabotage-, and ultimately nuclear EMP-attack to impair the United States

quickly and decisively by blacking-out large portions of its electric grid and other critical infrastructures. Foreign adversaries may aptly consider nuclear EMP attack a weapon that can gravely damage the U.S. by striking at its technological Achilles Heel, without having to confront the U.S. military.[60]

The report indicates that 'Russia, China, and North Korea now have the capability to conduct a nuclear EMP attack against the United States. All have practiced or described contingency plans to do so.'[61] According to congressional testimony given by William Graham (chair of the EMP Commission), the consequence of a major EMP attack could be the death of 90 percent of all Americans, as the infrastructure could not be quickly rebuilt and as only 10 percent of the current U.S. population could be sustained with 18th-century technology.[62]

Nuclear EMPs

The most serious EMP threat would be a HEMP detonated over the center of the United States, which could disrupt the electricity grid across the United States and, hence, could potentially impact the United States' ability to identify and retaliate against the attacker. According to Pry,

> [a]ny nuclear weapon detonated at an altitude of 30 kilometers or higher will generate a potentially catastrophic HEMP . . . a ten kiloton weapon . . . would generate an HEMP field about 600 kilometers in radius . . . the entire Eastern grid would certainly be plunged into a protracted blackout.[63]

Pry has claimed that HEMPs fall into the category of NPP weapons and that they could be part of a new RMA.

An article published in 2011 in the journal of the Russian General Staff, *Military Thought*, referenced the use of '[a] single low-yield exploded . . . high above the area of combat operations can generate an electromagnetic pulse covering a large area and destroying electronic equipment without loss of life that is caused by the blast or radiation.'[64] The article also mentioned that a HEMP could be an effective anti-satellite weapon to knock out military and civilian communications.[65] Not only has Russia pondered the strategic utility of HEMPs, but it has already made a veiled threat to use a HEMP in relation to Russia's opposition to the 1999 NATO Kosovo War.[66] Similarly, China has also considered the strategic benefits of HEMP weapons as they have the ability to degrade 'informatized' military equipment.[67] Even North Korea has the ability to carry out a HEMP attack against the U.S. mainland with the help of its KMS-3 and KMS-4 satellites, which pass over the United States daily and which could carry a miniaturized nuclear warhead.[68]

Owing to several balloons or objects that were shot down by the U.S. Air Force over U.S. territory in February 2023, there have been increasing concerns that an adversary could deploy a nuclear EMP weapon with a high-altitude balloon.[69]

A Chinese high-altitude balloon traversed the United States from Alaska to the coast of South Carolina from February 1 to February 4, flying at an altitude of about 60,000 feet, or 18 kilometers above the surface, and demonstrating an ability to maneuver.[70] Previously, in 2018, China seemed to have tested a high-altitude balloon as a launch platform for three hypersonic glide vehicles, indicating that balloons are being considered for use as weapons platforms.[71]

Balloons are an old technology and they have many drawbacks, such as slow speed, poor maneuverability, and poor survivability once detected. However, they also have some characteristics that could make them preferrable to a long-range missile or satellite attack with an EMP weapon: The Chinese spy balloon was difficult to detect (detection occurred only by chance when a passenger in an airliner spotted the unusual object), and this could enable strategic surprise; in addition, it could be difficult to determine where the attack came from (a balloon could be launched from a ship in the middle of the ocean) and who the attacker was (interestingly, the Chinese balloon used U.S. commercial materials and equipment to make attribution difficult).[72]

Non-nuclear EMPs

An EMP can be generated without a nuclear detonation, using so-called radiofrequency (RF) or high-power microwave (HPM) weapons, which could produce electromagnetic radiation strong enough to disrupt or destroy electronics. HPMs are like 'electronic warfare on steroids': 'HPM can blast through any filter and overpower the electronics to make them a pulpy metallic mess,' according to Doug Beason.[73] A demonstration conducted at the U.S. Army Aberdeen Test Center in 2001 showed that a radiofrequency generator could interfere with commercial electronics, and that electronic components readily available in Radio Shack could be built into a non-nuclear EMP (NNEMP) weapon.[74]

The U.S. military developed an NNEMP weapon called the Counter-electronics High Power Microwave Advanced Missile Project (CHAMP); a prototype was tested in 2012, and, in 2019, there were at least 20 operational CHAMP missiles ready to be launched from B-52 bombers that could be used against North Korea or Iran.[75] China has reportedly developed a hypersonic missile that can travel 3,000 kilometers at Mach 6 to deliver an NNEMP warhead that will destroy electronics within a two-kilometer radius of the target.[76] Obviously, hypersonic missiles cannot be intercepted and, hence, represent a very severe threat to U.S. national security, especially if they carry an NNEMP-type warhead.

Pry suggests that most NNEMP generators only have a maximum range of ten kilometers, meaning that they can only cause localized damage to electronics.[77] However, they could be deployed on the ground in a vehicle in strategic locations such as Wall Street or near a substation or transformer to disrupt key elements of an electric grid. The weapon could be small enough to fit into a suitcase and be carried by a single person. There is even a U.S. company that sells a 62-lb EMP device in a suitcase for industrial purposes.[78] According to Pry, '[d]ozens of nations reportedly have NNEMP weapons or are developing them,' which would include 'Russia,

China, North Korea, Iran, Pakistan, India, Israel, Germany, the United Kingdom, France, Australia, and Switzerland.'[79] Pry has claimed that

> North Korea used a non-nuclear EMP weapon, a 'radiofrequency cannon' purchased from Russia, to threaten and stop air traffic flying into Seoul, South Korea's capital, in December 2010; March 9, 2011; and April–May 2012. The attacks caused widespread communications blackouts and pre-vented automobiles from starting in South Korean communities along the Demilitarized Zone.[80]

As pointed out by the EMP Commission, NNEMPs could be used in conjunction with other approaches for greater reliability and impact: 'Military history and common sense suggests that a threefold attack—using cyber, sabotage, and EMP—will be better than an attack using just one of these.'[81] NNEMP devices could be easily smuggled into the United States and strategically deployed at important targets, coordinating with cyberattacks and other physical sabotage to inflict maximum damage to an electric grid. Deliberately caused blackouts could also be used to facilitate other terrorism or high-end crime or to cause societal chaos due to rioting and looting.

A less sophisticated attack could be carried out by domestic or foreign terror groups using small NNEMP devices transported in vans that are driven around to sequentially attack a number of substations. In a scenario developed by Pry, a single NNEMP truck could attack 30 extra-high-voltage transformer control sub-stations in 24 hours if the truck stopped for ten minutes to irradiate each target and then drove to the next target, 40–50 miles away.[82] An adversary would only need to destroy 9 out of 2,000 substations to cause a protracted nationwide blackout.[83]

Environmental Modification Weapons

The idea of using a modified environment as a weapon goes back a long time. When Leonardo da Vinci was painting the *Mona Lisa*, he was also working on a project that concerned the background landscape of the world's most iconic painting. The project, which ultimately failed, was the planned diversion of the Arno river to de-prive Florence's rival, Pisa, of freshwater in the Second Italian War (1499–1504).[84] The specter of environmental modification loomed large in the early Cold War as nuclear weapons enabled environmental modification on a global scale for the first time. There was concern that the Soviets might try to melt the polar ice caps with nuclear weapons to flood major cities in Western Europe and the United States.[85] In 1977, the superpowers signed the Environmental Modification Convention, which paradoxically acknowledged the possibility of environmental warfare while not actually banning anything owing to the vagueness of the treaty.[86] The area of environmental warfare has been extremely secretive and it remains a concern, as some countries may be willing and able to use environmental modification as a means to harm others.

Established Techniques

Environmental modification as a means of war is not new and, at a minimum, goes back to the Second World War when Germany and the Allies directed military attacks against dikes and dams in order to cause flooding.[87] The main established techniques of environmental warfare are the use of incendiaries to cause large fires, the release of radiation to make areas uninhabitable, the deliberate infection of plants and livestock to attack food production, and weather manipulation through cloud seeding.

Fire

Uncontrolled fires are enormously destructive, and fire can be a very powerful weapon of war. The damage caused by the firebombing of Tokyo exceeded the damage caused by all other bombing in the war, including the two nuclear bombs.[88] According to the Stockholm International Peace Research Institute (SIPRI), 'the destruction of vegetation by incendiary means can be used to deny the enemy forest cover, food, feed or industrial crops of one sort and another.'[89] Between 1965 and 1967, the U.S. military attempted to start massive forest fires in Vietnam, one of which was Operation Sherwood Forest, resulting in the destruction of 116 square miles of forest in the Tây Ninh province.[90] During the 1991 Gulf War, Iraq engaged in environmental warfare by deliberately setting 500 Kuwaiti oil wells on fire, which burned 400,000 tons of crude oil and blackened the sky with smoke for months.[91] A 20°C drop in temperature was reported within 50 kilometers of the fires.[92] The likely motivations were to destroy Kuwaiti crude oil reserves before the Iraqi withdrawal and to provide a smokescreen for Iraqi troops leaving Kuwait, as satellites and airborne reconnaissance cannot see through dense smoke. Many of the total of 1,200 fires may have been caused by Allied bombing.[93] Fire is also a great tool for sabotage short of war and is suitable for use against certain industrial or area targets.

In today's world, spaceborne or airborne high-energy lasers could be used to ignite targets from afar—for example, to cause forest fires or fires that target key infrastructure and human settlements. The U.S. military worked on spaceborne lasers in the context of the Strategic Defense Initiative in the 1980s, and so did the Soviet military. The Soviets even developed and launched a spaceborne laser called Skif-DM in 1987, which was unsuccessful.[94] Since the Cold War, the U.S. Air Force has worked on an airborne laser based on a modified 747 and meant to shoot down enemy missiles in the boost phase from a distance of hundreds of miles.[95] After successfully testing an airborne tactical laser on a C-130 Hercules in 2009, the U.S. Air Force has recently received its first tactical laser that can be mounted on a fighter jet, while ground-based lasers are also under development.[96]

The U.S. Department of Agriculture discussed the development of a truck-mounted laser ignition device that could be used for forest management, suggesting that '[t]he controlled firing techniques aim to mimic nature's wildfire which originally produced much of the existing forests.'[97] In short, lasers might target dry patches of forest to ignite them and allow controlled burns of vulnerable forests. If one was to use a high-energy laser for the purpose of covert sabotage, the primary advantage would be that it leaves behind no evidence in the form of traces of incendiaries. It is not a huge stretch to suggest that spaceborne or airborne lasers could be used by an adversary to deliberately start massive wildfires that cause tremendous strains on a country's power grid, economy, and the health of the population (owing to the massive amount of smoke).

Radiation

An obvious method of causing large-scale environmental damage is the deliberate release of harmful radiation to affect human health, damage the food supply, and inflict severe economic damage. Before the feasibility of nuclear fission weapons was experimentally proven, the U.S. Army considered radiological weapons, to be used against area targets such as cities, a fallback option. More bizarrely, there was even a proposal by Manhattan Project scientists to put strontium in the milk supply to poison and kill people within two months as a result of damaged bone marrow.[98] The U.S. government also feared that the Germans might use radiation against the Allied forces in Normandy and, hence, provided the landing forces with Geiger counters.[99]

In 1952, the U.S. Army developed a radiological bomb codenamed E-83, which dispersed 70 lb of the isotope tantalum-181 and was tested at Dugway Proving Ground, Utah.[100] Both superpowers investigated 'salted nuclear weapons' that could maximize the amount of long-lasting radioactive fallout to make large parts of a country uninhabitable.[101] A cobalt-60 bomb, first proposed by Leo Szilard, became the 'doomsday weapon' of popular culture owing to its long half-life combined with its ability to generate deadly gamma radiation.[102] Both superpowers eventually abandoned the concept of radiological weapons, mostly because of their low cost-effectiveness and limited military utility.

This is not to suggest that radiological weapons (RWs) and radiological warfare would be irrelevant in today's world. According to Samuel Meyer, Sarah Bidgood, and William Potter,

> [n]otwithstanding the failure to date of any country to incorporate radiological weapons into its military arsenal . . . [the] pursuit of RW is far more widespread than is commonly known. Although some of the factors that inhibited the adoption of RW in the past may still pertain, one cannot exclude the possibility that some countries today, such as North Korea, may be exploring an RW option or may choose to do so in the near future.[103]

What has not received much attention is the possibility of weaponizing radiation in a more covert or potentially deniable manner. This includes the deliberate sabotage of nuclear reactors and facilities, but also terrorist dirty bombs, as well as the covert introduction of radiation into the water and food supplies.

The Chernobyl accident of April 1986 caused the release of 3 or 4 percent of the nuclear fuel, which, while failing to cause many immediate deaths, did result in much anxiety and likely increased the cancer risk of the local population.[104] The Fukushima nuclear accident of March 2011 was substantially worse than the Chernobyl accident. In 2011, the Japanese government, which was unable to capture all the radioactive water, approved a plan to release 1 million tons of it into the Pacific over a 30-year period.[105] A state or nonstate actor might deliberately sabotage a nuclear reactor in order to escalate a conflict or create a crisis in an adversarial country. Since the beginning of the Russia–Ukraine War, there have been concerns about the potential sabotage of a Ukrainian nuclear power plant by one of the belligerents in order to inflict more distress and chaos.[106]

Agroterrorism

The environmental modification threat includes the deliberate modification of biological systems, especially in relation to agriculture and livestock. Simon Whitby, Piers Millett, and Malcolm Dando have argued that '[h]istorically, it is clear that anti-animal and anti-plant biological warfare was viewed predominantly as a potential strategic-level operation aimed at large-scale economic and social disruption and disintegration,' and that 'this will remain a major aim if biological warfare cannot be prevented.'[107] Possibilities here are the release of biological agents into the environment to infect certain plants and animals, the genetic engineering of organisms that are sterile, and the use of gene drives to displace natural organisms.

Agroterrorism is a viable mode for a bioterrorism attack aimed at disrupting the food supply and causing economic damage and has been explored by a number of states as a method of covert economic attack. During the First World War, Germany attacked Allied packhorses and mules with glanders and anthrax with the help of saboteurs, and the French may have attacked German livestock with glanders and anthrax.[108] The U.S. government also pondered biological attacks against crops during the early Cold War period. For example, the U.S. Joint Chiefs of Staff recommended Operation Square Dance to President Kennedy, which was about the covert dispersal of a plant parasite called Bunga on Cuba, which was meant to destroy sugar cane plants, Cuba's primary cash crop.[109] Under Saddam Hussein, Iraq developed 'infectants that make wheat unfit for human consumption.'[110]

David Malet argued that

> [a]n outbreak of foot-and-mouth disease (FMD) decimated the British agriculture and tourist industry in 2001, and the importation of a single pig from Hong Kong infected with FMD caused the loss of $19 billion to the Taiwanese pork industry in livestock that required destruction.[111]

According to Michael Petersen, '[f]oot-and-mouth disease, one of the most contagious viruses known, might be the perfect weapon for an agroterrorism attack.'[112] The virus is harmless to humans and, hence, risk-free for terrorists, highly contagious for livestock, easy to distribute due to lack of security on farms, and less likely to cause a backlash for the terrorists as humans are not harmed.[113] A further challenge is that biological threats are notoriously difficult to evaluate, which leads governments to overreact to biological threats with mass destruction of livestock that may have been at risk. The 2022 avian flu outbreak led to the culling of over 140 million birds in the United States, United Kingdom, and European Union, which represents an economic loss of close to $1 billion.[114]

Weather Manipulation

The weather and climate have an enormous influence on many aspects of human activities, especially in relation to agriculture, transportation, and military operations. Drought and rainfall can destroy crops, flooding can destroy roads and other infrastructure, fog can impact air traffic and military ground operations, and severe weather events such as extreme cold or heat can place enormous strain on electricity grids, causing blackouts. The ability to influence the weather has obvious military utility. A well-known Air Force research paper from 1996 stated:

> A global, precise, real-time, robust, systematic weather-modification capability would provide war-fighting CINCs with a powerful force multiplier to achieve military objectives. Since weather will be common to all possible futures, a weather-modification capability would be universally applicable and have utility across the entire spectrum of conflict. The capability of influencing the weather even on a small scale could change it from a force degrader to a force multiplier.[115]

Military research into weather modification goes back to the 1940s. In 1946, General Electric discovered that clouds could be seeded by dropping dry ice from planes.[116] The British Ministry of Defence conducted Project Cumulus from 1949 to 1952 in an effort to seed clouds, which some believe resulted in the accidental flooding of the village of Lynmouth, Devon, in August 1952.[117] In 1962, the U.S. Navy attempted to create and steer hurricanes in Project Stormfury, which involved cloud seeding in the path of a tropical storm with the aim of dissipating massive hurricanes before they reached landfall.[118] Also widely known is Project Popeye, which attempted to flood the Ho Chi Minh trail from 1967 to 1972 by seeding clouds with silver and lead iodide in a bid to disrupt the Vietcong supply lines, aid U.S. bombing missions, and disrupt enemy offensives with bad weather.[119] Rain-making through cloud seeding is a well-established approach to weather modification that can affect local weather but can also affect weather elsewhere, as extended precipitation in one area may cause drought in another area.[120] It is also widely known that Russia and China have conducted a lot of research into weather modification and that they routinely use cloud seeding to mitigate droughts and wild fires and to clear the skies for parades on national holidays.[121]

Future Geophysical Weapons

Geophysical weapons are speculative and are mostly referred to in the Russian literature, with scarce information in Western literature. Geophysical weapons could trigger major environmental catastrophes such as earthquakes, tsunamis, volcano eruptions, global cooling, or damage to the ozone layer. Research in this field tends to be unacknowledged and may be conducted in almost complete secrecy. While it is impossible to determine whether such weapons exist, there are some indications that they are scientifically feasible.

Tectonic Weapons

Before the end of the Second World War, the United States and New Zealand conducted 4,000 test explosions for the development of a 'tsunami bomb' in Project Seal, which could destroy coastal cities with tidal waves. The project was terminated just months before the nuclear bombing of Hiroshima and Nagasaki. Although no tsunami was ever triggered, the final report of Project Seal indicated that the scientists deemed it feasible to create tsunamis with underwater explosions.[122] The Soviet Union began the development of so-called tectonic weapons in 1954 with the systematic study of the effects of underground nuclear explosions. According to Mary FitzGerald,

> [t]hese are weapons that generate natural catastrophes such as earthquakes, torrential rains, tsunamis, and destruction of the ozone layer. It is possible to trigger earthquakes with underground explosions of powerful nuclear charges, particularly in areas of high seismic activity. It is also possible to trigger tsunamis with an explosion of nuclear charges in certain areas of seas and oceans.[123]

Timothy Thomas has suggested that '[t]rigger mechanisms [for geophysical weapons] could, for example, influence geophysical stress points, such as the junction of tectonic plate movements where a significant amount of energy is accumulating.'[124]

In 1987, the Central Committee of the Communist Party of the Soviet Union initiated a secret project codenamed Mercury, which was intended, according to declassified documents, to 'develop a methodology for remote operation on an earthquake epicentre by using weak seismic fields and research possibilities of transferring the seismic energy of an explosion.'[125] The research was carried out in Azerbaijan, and initial tests were carried out in 1990 and were deemed to have been successful.[126] The research was disrupted by the collapse of the Soviet Union, as Azerbaijan became a sovereign country. Russia started a new research program, codenamed Volcano, in 1992, which conducted tests involving underground nuclear explosions, possibly at Novaya Zemlya, a nuclear test region in Northern Russia, in 1992 and 1993.[127]

Little public information is available on the current status of Russian tectonic weapons research other than vague acknowledgments that research into geophysical weapons would continue in Russia.[128] No hard evidence for the existence of tectonic weapons has been presented, which does not rule out the possibility that such weapons could be used covertly or could be developed in the future.

Ionospheric Heaters

In 1985, Bernard Eastlund filed a patent titled 'A Method and Apparatus for Altering a Region of the Earth's Atmosphere, Ionosphere, and Magnetosphere,' which is what is now described as an ionospheric heater as it beams electromagnetic radiation into an upper layer of the atmosphere.[129] The patent stated '[w]eather modification is possible by, for example, altering upper atmosphere wind patterns or altering solar absorption patterns by constructing one or more plumes of atmospheric particles which will act as a lens or focusing device.'[130]

In 1993, the U.S. Air Force announced the award of a contract for the construction of the High-frequency Active Auroral Research Project (HAARP) to APTI, which was a subsidiary of ARCO and had acquired the Eastlund patents.[131] The site chosen was Gakona, Alaska, which now houses a large array of 180 antennas. The Department of Defense spent over $300 million on HAARP, which was only completed in 2007 and was transferred to the University of Alaska, Fairbanks, in 2015.[132]

The stated purpose of HAARP is ionospheric research, but suspicions about the true purpose and capabilities of HAARP have endured. The fact that the U.S. Air Force Research Laboratory (AFRL) funded the project for over 20 years would indicate that it had some military application in mind in relation to heating the ionosphere.[133] A report to the European Parliament from 1998 speculated that '[t]he aim of HAARP is to control and manipulate the ionosphere so as to enable the manipulator to wipe out communications at will on a global scale, or to make them resilient in the event of a nuclear war.'[134] A Russian military journal similarly argued that

> [t]he layer of ionosphere which is excited by HAARP influences the radio and electronic equipment which is installed in the military hardware: Fire control and guidance systems, fire adjustment equipment, navigation systems, etc. As a result, an aircraft or a missile will be damaged if they fly through the beam.[135]

In this context, it is certainly a concern that China and Russia jointly conducted a major experiment in the ionosphere with an ionospheric heater located in Vasilsursk, Russia, in late 2018. According to Peter Dockrill, '[b]y selectively disturbing the charged particles that make up this part of the upper atmosphere, scientists or even governments could theoretically boost or block long-range radio signals.'[136] The test caused a physical disturbance over an area of 49,000 square miles.[137] China has also built an ionospheric heater located on an island in Hainan Province, which could be used to jam radio or GPS signals over the South China Sea. According to Stephen Chen, '[t]he militaries have been in a race to control the ionosphere for decades . . . Changing the ionosphere over enemy territory can also disrupt or cut off their communication with satellites.'[138] There are currently eight active ionospheric heaters worldwide: HAARP (Gaukona, AK), SuperDARN (Jicamarca, Peru), Arecibo Observatory (Puerto Rico), Tromsø Ionospheric Heater (Norway), Sura Ionospheric Heater (Russia), MST Radar Facility (Gadanki, India), Shigaraki MU Observatory (Shigaraki, Japan), and Hainan Coherent Scatter Phased Array (Sanya, China).

Atmospheric Geoengineering

Geoengineering was proposed as a solution to climate change at the climate workshop at Harvard University in late 2007 by scientists such as David Keith, who is a major advocate for cooling the global climate.[139] The Obama administration sponsored research related to geoengineering with 'ideas like setting up sun shields in space or dispersing microscopic particles in the air to make coastal clouds more reflective, dissipate heat-trapping cirrus clouds, or scatter sunlight in the stratosphere.'[140] Proponents of atmospheric geoengineering want to mitigate the negative effects of carbon dioxide emissions and thereby slow down global warming.[141] The administration even considered climate change a major threat to national security and tasked the military and intelligence services with monitoring the threat. From 2010 to 2012, the CIA had an office for monitoring global climate change.[142] In 2015, the CIA gave $600,000 to fund a report by the National Academy of Sciences on climate change, which could suggest that the CIA is interested in military applications of geoengineering technologies, perhaps used by adversaries against the United States.[143]

Atmospheric geoengineering received more extensive news coverage in 2022 and 2023, which might suggest that governments were seeking public approval for the manipulation of the climate on a planetary scale. In particular, stratospheric aerosol injection is being discussed as a method for reflecting sunlight back into space in order to reduce global warming. Scientists proposed the creation of an artificial volcano by dispersing 1.5 million tons of sulphate particles in the stratosphere to dim the sun and cool the planet.[144] Balloons or special aircraft would release reflective particles such as sulfur dioxide into the upper atmosphere, which would block the sunlight. The ideas moved to the stage of real-world testing, with high-altitude balloon tests for the dispersal of particles being carried out in New Mexico.[145]

Obviously, the idea is highly controversial owing to the risk of unintended effects such as distorting established weather and climate patterns, with potential damage to crops, and it could damage or harm the recovery of the ozone layer, causing more harmful ultraviolet rays to penetrate the atmosphere.[146]

The big political issues would be: Who would implement such a plan to manage the global climate, and who would decide which regions should get the greater benefits and which ones should experience the more negative effects? The issue is more urgent than it seems. In January 2023, the Mexican government decided to prohibit any geoengineering experiments within its country after an American start-up company, Make Sunsets, carried out atmospheric experiments without asking for permission from the Mexican Ministry for the Environment and Natural Resources.[147] Furthermore, if climate engineering was weaponized, it could be used to darken the skies over adversarial countries to cause crop failures or could induce droughts or flooding by changing weather patterns, resulting in food shortages or starvation.

Conclusion

Sixth generation warfare can be described as the weaponization of technical systems and even nature against a population in order to cause societal disruption, economic ruin, or mass starvation. The respective technologies and approaches have been already investigated during the Cold War. Modern societies have become even more dependent on technical systems, making deliberate attacks against them so much more effective. Cyberattacks, EMP attacks, and environmental modification can knock out electricity grids, communications, and transportation systems and impact the availability of energy and food, all of which are critically important to sustain human civilization. Of course, in a globalized world, everybody would be affected if the lights did go out for more than a few days in the United States or in China. Sixth generation warfare could be a losing proposition for everyone, since everyone is vulnerable to disruption of their technical systems and the global supply chain. However, in 6GW, the primary advantage will lie in building resilience and in the study of interdependencies in order to produce effects that largely hurt the adversary rather than one's own side. In a covert war for global domination, it could still be worth it to suffer the temporarily negative consequences of global disruption in order to emerge more powerful on the other side of the disruption.

Notes

1 Roger McDermott has suggested that Russian military theory distinguishes between five generations of warfare in military history: '[T]he first saw edged weapons; the second, gunpowder-based; third, rifled weapons; fourth, automatic weapons including industrialization of warfare, and the fifth, nuclear.' Roger McDermott, 'Russia's Entry to Sixth-Generation Warfare: The "Non-Contact" Experiment in Syria,' Jamestown Foundation, May 29, 2021, available at: https://jamestown.org/program/russias-entry-to-sixth-generation-warfare-the-non-contact-experiment-in-syria/
2 Mary FitzGerald, 'The Russian Military's Strategy for "Sixth Generation" Warfare,' *Orbis* (Summer 1994): 457–458.
3 Stephen Blank, '"Putin's Asymmetric Strategy": Nuclear and New-Type Weapons in Russian Defense Policy,' in: Glen E. Howard and Matthew Czekaj (eds.), *Russia's Military Strategy and Doctrine* (Washington, DC: Jamestown Foundation Press, 2019), 257.

4 Peter Vincent Pry, *Blackout Warfare: Attacking the U.S. Electric Grid, a Revolution in Military Affairs* (EMP Task Force on National and Homeland Security, 2021), 121.
5 Patrick Dugan, 'Sunsets and Dawns: The End-Game for 5GW and the Human Era,' in: Abbott (ed.), *The Handbook of 5GW*, 103.
6 PakDefense.com, '6th Generation of Warfare: Evolution of Warfare from Classic to Hybrid Domains,' PakDefense.com, March 28, 2023, available at: www.pakdefense.com/blog/6th-generation-warfare-evolution-of-warfare-from-classic-to-hybrid-domains/, accessed July 4, 2023.
7 Adam Herring, 'Searching for 5GW,' in: Abbott (ed.), *The Handbook of 5GW*, 71.
8 Stephen Pampinella, 'The Construction of 5GW,' in: Abbott (ed.), *The Handbook of 5GW*, 50.
9 Dugan, 'Sunsets and Dawns,' 105.
10 *Britannica Dictionary*, 'Sabotage,' available at: www.britannica.com/dictionary/sabotage, accessed July 4, 2023.
11 Thomas K. Adams, *US Special Operations Forces in Action: The Challenge of Unconventional Warfare* (London: Frank Cass, 1998), XXIV.
12 Martin Libicki, 'The Specter of Non-obvious Warfare,' *Strategic Studies Quarterly* (Fall 2012): 96.
13 Libicki, 'The Specter of Non-obvious Warfare,' 96.
14 Bernd Horn, James D. Kiras, and Emily Spencer (eds), *The (In)visible Hand: Strategic Sabotage Case Studies* (Ottawa: Canadian Special Forces Command, 2021), 4.
15 Christopher Andrew and Vasili Mitrokhin, *The Mitrokhin Archive: The KGB in Europe and the West* (London: Penguin Books, 1999), 468.
16 Stanislav Lunev, *Through the Eyes of the Enemies: Russia's Highest Ranking Military Defector Reveals Why Russia Is More Dangerous than Ever* (Washington, DC: Regnery, 1998), 24.
17 Lunev, *Through the Eyes of the Enemies*, 25; Richard C. Paddock, 'Lebed Says Russia Has Lost Track of 100 Nuclear Bombs,' *Los Angeles Times*, September 9, 1997.
18 John Psaropoulos, 'Europe Awakens to the Threat of Sabotage by Russian Agents,' Al Jazeera, January 17, 2023, available at: www.aljazeera.com/news/2023/1/17/europe-awakens-to-the-threat-of-sabotage-by-russian-agents
19 According to an estimate from the World Economic Forum, a global internet outage would cost at least $50 billion per day and over $1 trillion after three weeks. World Economic Forum, 'What the COVID-19 Pandemic Teaches Us about Cybersecurity—and How to Prepare for the Inevitable Global Cyberattack,' World Economic Forum, June 1, 2020, available at: www.weforum.org/agenda/2020/06/covid-19-pandemic-teaches-us-about-cybersecurity-cyberattack-cyber-pandemic-risk-virus/
20 Mary Manjikian, *Introduction to Cyber Politics and Policy* (Los Angeles, CA: Sage, 2021), 1–2.
21 Reuters, 'Massive Network Outage in Canada Hits Homes, ATMs and 911 Emergency Lines,' *The Guardian*, July 8, 2022, available at: www.theguardian.com/world/2022/jul/08/internet-down-canada-rogers-mobile-network-outage
22 U.S. Energy Information Administration, 'U.S. Energy Facts Explained,' available at: www.eia.gov/energyexplained/us-energy-facts/, last updated June 10, 2022.
23 Notable here is the sabotage of the Nord Stream gas pipelines connecting Russia and Germany in September 2022, which resulted in a serious energy shortage in Germany and also caused an environmental disaster by releasing 14.6 million tons of greenhouse gases into the atmosphere.
24 Pry, *Blackout Warfare*, 12.
25 Pry, *Blackout Warfare*.
26 Trucks move over 70% of all goods in the United States. For years there have been concerns about diesel engine oil shortages that could disrupt the trucking industry and logistics, with supermarkets and stores no longer getting the necessary deliveries to keep products stocked.

27 Ore Koren, Benjamin Bagozzi, and Thomas Benson, 'Food and Water Insecurity as Causes of Social Unrest: Evidence from Geolocated Twitter Data,' *Journal of Peace Research* 58, no. 1 (2021): 67.

28 The Third Reich developed a Hunger Plan after the German invasion of the Soviet Union in June 1941 to starve 20–30 million Soviets in order to facilitate the conquest and occupation of the Western Soviet Union. Alex de Vaal, *Mass Starvation: The History and Future of Famine* (Cambridge: Polity Press, 2018), 101–103.

29 Henry Kissinger infamously commissioned a National Security Council study in 1973 (NSSM 200) to investigate how the United States can geopolitically exploit the dependency of other countries on U.S. food exports.

30 Matt Ford, 'Why Doesn't the Government Stockpile Other Essentials as It Does Oil?' *The New Republic*, November 24, 2021, available at: https://newrepublic.com/article/164537/joe-biden-gas-national-reserve

31 Libicki, 'The Specter of Non-obvious Warfare,' 92.

32 Thomas Rid, *Cyber War Will Not Take Place* (Oxford: Oxford University Press, 2013), 7.

33 Kim Zetter, *Countdown to Zero Day: Stuxnet and the Launch of the World's First Digital Weapon* (New York: Crown, 2014), 64.

34 WikiLeaks, 'Vault 7: CIA Hacking Tools Revealed,' Press Release, March 7, 2017, available at: https://wikileaks.org/ciav7p1/, accessed July 4, 2023.

35 Fred Kaplan, *Dark Territory: The Secret History of Cyber War* (New York: Simon & Schuster, 2016), 167–168.

36 Jon R. Lindsay, 'Stuxnet and the Limits of Cyber Warfare,' *Security Studies* 22, no. 3 (2013): 384.

37 Lindsay, 'Stuxnet and the Limits of Cyber Warfare,' 381–382.

38 Christopher Bronk and Eneken Tikk-Ringas, 'The Cyber Attack on Saudi Aramco,' *Survival* 55, no. 2 (2013): 81–96.

39 David Sanger, Clifford Krauss, and Nicole Perlroth, 'Cyberattack Forces Shutdown of a Top U.S. Pipeline,' *New York Times*, May 8, 2021.

40 Lauren Fedor, Myles McCormick, and Hannah Murphy, 'Biden Says "Strong Reason" to Believe Pipeline Hackers Are in Russia,' *Financial Times*, May 13, 2021.

41 Jill Leovy, 'Cyberattack Cost Maersk as Much as $300 Million and Disrupted Operations for Two Weeks,' *Los Angeles Times*, August 17, 2023.

42 Carly Page, 'Maritime Giant DNV Says 1,000 Ships Affected by Ransomware Attack,' *TechCrunch*, January 18, 2023, available at: https://techcrunch.com/2023/01/18/dnv-norway-shipping-ransomware/

43 Josh Margolin, Sam Sweeney, and Quinn Owen, 'Cyberattacks Reported at US Airports,' *ABC News*, October 10, 2022, available at: https://abcnews.go.com/Technology/cyberattacks-reported-us-airports/story?id=91287965

44 Pete Muntean and Gregory Wallace, 'FAA System Outage Causes Thousands of Flight Delays and Cancellations across the US,' CNN, January 11, 2023, available at: www.cnn.com/travel/article/faa-computer-outage-flights-grounded/index.html

45 FBI, 'Ransomware Attacks on Agricultural Cooperatives Potentially Timed to Critical Seasons,' Private Industry Notification, April 20, 2022, available at: www.ic3.gov/Media/News/2022/220420-2.pdf

46 Sean Lyngaas, 'Cyberattack on Food Giant Dole Temporarily Shuts Down North America Production, Company Memo Says,' CNN, February 22, 2023, available at: www.cnn.com/2023/02/22/business/dole-cyberattack/index.html

47 Richard Clarke and Robert Knake, *Cyber War: The Next Threat to National Security and What to Do About It* (New York: Harper Collins, 2010), 167.

48 National Academies of Sciences, *Terrorism and the Electric Power Delivery System* (Washington, DC: National Academies Press, 2012), 2.

49 Damien van Puyvelde and Aaron Brantley, *Cybersecurity: Politics, Governance and Conflict in Cyberspace* (Cambridge: Polity, 2019), 99.

50 Ted Koppel, *Lights Out: A Cyberattack, a Nation Unprepared Surviving the Aftermath* (New York: Broadway Books, 2015), 95.
51 Adam Xu, 'US Moves to Pull Chinese Equipment from Its Power Grid,' *Voice of America*, May 9, 2020, available at: www.voanews.com/a/east-asia-pacific_voa-news-china_us-moves-pull-chinese-equipment-its-power-grid/6188992.html
52 David Sanger and Mark Mazzetti, 'U.S. Had Cyberattack Plan if Iran Nuclear Dispute Led to Conflict,' *New York Times*, February 16, 2016.
53 Kalev Leetaru, 'Could Venezuela's Power Outage Really Be a Cyber Attack?' *Forbes*, March 9, 2019, available at: www.forbes.com/sites/kalevleetaru/2019/03/09/could-venezuelas-power-outage-really-be-a-cyber-attack/?sh=8fd7b7c607c7
54 David Sanger and Nicole Perlroth, 'U.S. Escalates Online Attacks on Russia's Power Grid,' *New York Times*, June 15, 2019.
55 Nicole Sganga, 'Physical Attacks on Power Grid Rose by 71% Last Year, Compared to 2021,' *CBS News*, February 22, 2023, available at: www.cbsnews.com/news/physical-attacks-on-power-grid-rose-by-71-last-year-compared-to-2021/
56 Jacob Darwin Hamblin, *Arming Mother Nature: The Birth of Catastrophic Environmentalism* (Oxford: Oxford University Press, 2013), 110.
57 Doug Beason, *The E-Bomb: How America's Directed Energy Weapons Will Change the Way Future Wars Will Be Fought* (Cambridge, MA: Da Capo Press, 2005), 107.
58 Pry, *Blackout Warfare*, 139.
59 EMP Commission, 'Report of the Commission to Assess the Threat to the United States from Electromagnetic Pulse (EMP) Attack,' EMP Commission Executive Report (2004), 1–2.
60 EMP Commission, 'Assessing the Threat from EMP Attack: Volume I,' Executive Report (July 2017), available at: https://apps.dtic.mil/sti/pdfs/AD1051492.pdf, 5.
61 EMP Commission, 'Assessing the Threat from EMP Attack,' 5.
62 William R. Graham, James Woolsey, and Peter Pry, 'Prepare for the Worst,' RealClear Defense.com, October 21, 2019, available at: www.realcleardefense.com/articles/2019/10/21/prepare_for_the_worst_114802.html
63 Pry, *Blackout Warfare*, 131, 133.
64 Pry, *Blackout Warfare*, 122.
65 A.V. Kopylov, 'Weak Points of the U.S. Concept of Network-Centric Warfare,' *Military Thought* 3 (2011), quoted from Pry, *Blackout Warfare*, 122.
66 Pry, *Blackout Warfare*, 123.
67 Peter Vincent Pry, 'China: EMP Threat,' EMP Task Force on National and Homeland Security, June 10, 2020, available at: https://info.publicintelligence.net/US-ChinaEMPThreat.pdf
68 Pry, *Blackout Warfare*, 125–128.
69 Paul Bedard, 'Dry Run: Balloons Called "Top Delivery Platform" for Nuclear EMP Attack,' *Washington Examiner*, February 3, 2023, available at: www.washingtonexaminer.com/news/washington-secrets/balloons-called-top-delivery-platform-for-nuclear-emp-attack
70 Sophie Bushwick, 'Chinese Spy Balloon Has Unexpected Maneuverability,' *Scientific American*, February 3, 2023, available at: www.scientificamerican.com/article/chinese-spy-balloon-has-unexpected-maneuverability/
71 Tyler Rogoway, 'Video Appears to Show China Testing Hypersonic Glide Vehicles via High Altitude Balloon,' *TheDrive.com*, September 21, 2018, available at: www.thedrive.com/the-war-zone/23758/video-appears-to-show-china-testing-hypersonic-glide-vehicles-via-high-altitude-balloon
72 Vivian Salama, Nancy Yousssef, and Michael R. Gordon, 'Chinese Spy Balloon Tracked over U.S. This Week; Incident Comes Days before Secretary of State Anthony Blinken's Trip to China,' *Wall Street Journal*, February 3, 2023; Nancy Youssef, 'Chinese Balloon Used American Tech to Spy on Americans; Preliminary Findings Show the Craft Collected Photos and Videos but Didn't Appear to Transmit Them, Officials Say,' *Wall Street Journal*, June 29, 2023.

73 Beason, *The E-Bomb*, 100.
74 Hunter Keeter, 'Non-Nuclear EMP Device Could Threaten Civil, Military Computer Systems,' *Defense Daily* 210, no. 23 (2001): 1.
75 John Keller, 'Air Force Deploys B-52 Missiles That Could Disable Enemy Military Electronics with High-Power Microwaves,' *Military Aerospace Electronics*, May 17. 2019, available at: www.militaryaerospace.com/rf-analog/article/14033453/air-force-deploys-b52-missiles-that-could-disable-enemy-military-electronics-with-highpower-microwaves
76 Dave Makichuk, 'China's EMP Missile Would Plunge Cities into Darkness,' *Asia Times*, September 29, 2021, available at: https://asiatimes.com/2021/09/chinas-emp-missile-would-plunge-cities-into-darkness/
77 Pry, *Blackout Warfare*, 101.
78 Pry, *Blackout Warfare*, 100.
79 Pry, *Blackout Warfare*, 106.
80 Peter Vincent Pry, 'Nuclear EMP Attack Scenarios and Combined-Arms Cyber Warfare,' Report to the Commission to Assess the Threat to the United States from Electromagnetic Pulse (EMP) Attack, July 2017, 10.
81 Pry, 'Nuclear EMP Attack Scenarios and Combined-Arms Cyber Warfare,' 20.
82 Pry, *Blackout Warfare*, 113.
83 Pry, *Blackout Warfare*, 114.
84 Roger D. Masters, *Fortune Is a River: Leonardo da Vinci's Magnificent Dream to Change the Course of Florentine History* (New York: Plume, 1999).
85 Hamblin, *Arming Mother Nature*, 131.
86 Hamblin, *Arming Mother Nature*, 215–216.
87 SIPRI, *Weapons of Mass Destruction and the Environment* (London: Routledge, 1977/2021), 88.
88 SIPRI, *Weapons of Mass Destruction and the Environment*, 84.
89 SIPRI, *Weapons of Mass Destruction and the Environment*, 83.
90 SIPRI, *Weapons of Mass Destruction and the Environment*, 86.
91 Abdullah Toukan, 'The Gulf War and the Environment: The Need for a Treaty Prohibiting Ecological Destruction as a Weapon of War,' *The Fletcher Forum of World Affairs* 15, no. 2 (1991): 97–98.
92 Rosalie Bertell, *Planet Earth: The Latest Weapon of War* (Dublin: Talma Studios, 2021), 49.
93 Bertell, *Planet Earth*, 50.
94 Dwayne A. Day and Robert Kennedy, 'Barbarian in Space: The Secret Space Laser Battle Station of the Cold War,' *The Space Review*, June 5, 2023, available at: <https://www.thespacereview.com/article/4598/1>.
95 Beason, *The E-Bomb*, 138–139.
96 Thomas Newdick, 'First Laser Weapon for a Fighter Delivered to the Air Force,' The Verge, July 22, 2022, available at: www.thedrive.com/the-war-zone/first-laser-weapon-for-a-fighter-delivered-to-the-air-force
97 Michael D. Waterworth, 'Laser Ignition Device and Its Application to Forestry, Fire and Land Management,' U.S. Department of Agriculture (1987), available at: www.fs.usda.gov/psw/publications/documents/psw_gtr101/psw_gtr101_01_waterworth.pdf, 182.
98 Barton J. Bernstein, 'Radiological Warfare: A Path Not Taken,' *Bulletin of Atomic Scientists* 41, no. 7 (1985): 46.
99 Bernstein, 'Radiological Warfare,' 47.
100 Joseph Trevithick, 'Exposed: The U.S. Army Tested Its Own Nuclear "Dirty Bomb,"' *National Interest*, October 14, 2015, available at: https://nationalinterest.org/blog/the-buzz/exposed-the-us-army-tested-its-own-nuclear-dirty-bombs%E2%80%99-14074
101 Samuel Meyer, Sarah Bidgood, and William C. Potter, 'Death Dust: The Little-Known Story of U.S. and Soviet Pursuit of Radiological Weapons,' *International Security* 45, no. 2 (2020): 51–94.

102 Edward Moore Geist, 'Would Russia's Undersea "Doomsday Drone" Carry a Cobalt Bomb?' *Bulletin of Atomic Scientists* 72, no. 4 (2016): 239.
103 Meyer et al., 'Death Dust,' 54.
104 Igor Khripunov, 'The Social and Psychological Impact of Radiological Terrorism,' *Nonproliferation Review* 13, no. 2 (2006): 287.
105 Yonglong Lu, Jingjing Yuan, Di Du, Bin Sun, and Xiaojie Yi, 'Monitoring Long-Term Ecological Impacts from Release of Fukushima Radiation into Ocean,' *Geography and Sustainability* 2 (2021): 95–96.
106 Matthew Bunn, 'The Largest Danger at the Zaporizhzhia Nuclear Power Plant: Intentional Sabotage,' *The Bulletin of Atomic Scientists*, July 6, 2023, available at: https://thebulletin.org/2023/07/the-largest-danger-at-the-zaporizhzhia-nuclear-power-plant-intentional-sabotage/
107 Simon Whitby, Piers Millett, and Malcom Dando, 'The Potential for Abuse of Genetics in Militarily Significant Biological Weapons,' *Medicine, Conflict and Survival* 18, no. 2 (2002): 144.
108 Jeanne Guillemin, *Biological Weapons: From the Invention of State-Sponsored Programs to Contemporary Bioterrorism* (New York: Columbia University Press, 2005), 21, 24.
109 Loch K. Johnson, *The Third Option: Covert Action and American Foreign Policy* (Oxford: Oxford University Press, 2022), 56–57.
110 David Malet, *Biotechnology and International Security* (Lanham, MD: Rowman & Littlefield, 2016), 110.
111 Malet, *Biotechnology and International Security*, 110.
112 Michael Petersen, 'Agroterrorism and Foot-and-Mouth Disease: Is the United States Prepared?' *Counterproliferation Papers Future Warfare Series* 13 (2002): IV.
113 Petersen, 'Agroterrorism and Foot-and-Mouth Disease,' 9.
114 Sophie Kevany, 'Avian Flu Has Led to the Killing of 140 Million Farmed Birds since Last October,' *The Guardian*, December 9, 2022, available at: www.theguardian.com/environment/2022/dec/09/avian-flu-has-led-to-the-killing-of-140m-farmed-birds-since-last-october
115 Tamzy J. House, James B. Near, William B. Shields, Ronald J. Celentano, David M. Husband, Ann E. Mercer, and James E. Pugh, 'Weather as a Force Multiplier: Owning the Weather by 2025,' Air Force 2025, August 1996, 3.
116 Bertell, *Planet Earth*, 114.
117 John Vidal and Helen Weinstein, 'RAF Rainmakers "Caused 1952 Flood,"' *The Guardian*, August 30, 2007, available at: www.theguardian.com/uk/2001/aug/30/sillyseason.physicalsciences
118 Hamblin, *Arming Mother Nature*, 201–202.
119 U.S. Congress, 'Weather Modification,' United States Senate, Subcommittee on Oceans and International Environment of the Committee on Foreign Relations, March 20. 1974.
120 SIPRI, *Weapons of Mass Destruction and the Environment*, 94.
121 Eva Hartog, 'Moscow Ready to Use Cloud Clearing Techniques on Victory Day,' *Moscow Times*, April 29, 2014, available at: www.themoscowtimes.com/2014/04/29/moscow-ready-to-use-cloud-clearing-techniques-on-victory-day-a34859; Reuters, 'Russia Is Seeding Rain Clouds in Siberia to Fight Wildfires That Have Been Burning across the Country for Weeks, Reducing Them to Just One Third of Their Former Size,' *Daily Mail*, July 10, 2020, available at: www.dailymail.co.uk/sciencetech/article-8510107/Russia-seeds-clouds-Siberia-douse-raging-wildfires.html; Helen Davidson, 'China "Modified" the Weather to Create Clear Skies for Political Celebration—Study,' *The Guardian*, December 5, 2021, available at: www.theguardian.com/world/2021/dec/06/china-modified-the-weather-to-create-clear-skies-for-political-celebration-study; and BBC, 'China Inducing Rainfall to Combat Severe Drought,' *BBC News*, August 17, 2022, available at: www.bbc.com/news/world-asia-china-62573547
122 Emily Bourke, 'Military Archives Show NZ and US Conducted Secret Tsunami Bomb Tests,' Australian Broadcasting Corporation, February 3, 2013.

123 Mary C. FitzGerald, *Impact of the RMA on Russian Military Affairs* (Washington, DC: Hudson Institute, 1997), 188.
124 Timothy Thomas, 'Advanced Weaponry and the Russian Art of War,' MITRE Center for Technology and National Security, October 2020, available at: www.mitre.org/sites/default/files/2021-11/prs-20-1890-russian-military-art-and-advanced-weaponry.pdf, 11.
125 Quoted from Carl Levitin, 'Russian Documents Set Out "Tectonic Weapon" Research,' *Nature* 383 (1996): 471.
126 FitzGerald, *Impact of the RMA on Russian Military Affairs*, 191.
127 Levitin, 'Russian Documents Set Out "Tectonic Weapon" Research,' 471.
128 Timothy Thomas, 'Russian Forecasts of Future of War,' *Military Review* (May–June 2019): 88.
129 Bernard Eastlund, 'Method and Apparatus for Altering a Region of the Earth's Atmosphere, Ionosphere, and Magnetosphere,' US4686605A, available at: https://patents.google.com/patent/US4686605A/en
130 Eastlund, 'A Method and Apparatus for Altering a Region of the Earth's Atmosphere, Ionosphere, and Magnetosphere.'
131 Jeanne Manning and Nick Begich, *Angels Don't Play This HAARP: Advances in Tesla Technology* (Anchorage, AK: Earthpulse Press, 1995), 28–29.
132 Sharon Weinberger, 'Alaska Military Site That Fueled Conspiracy Theories Will Be Transferred to Civilian Operators,' *The Intercept*, July 13, 2015, available at: https://theintercept.com/2015/07/13/military-site-center-conspiracy-will-transferred-civilian-operators/
133 The AFRL mission statement is: 'The Air Force Research Laboratory leads the discovery, development and integration of affordable warfighting technologies for our air, space and cyberspace forces.' See www.af.mil/About-Us/Fact-Sheets/Display/Article/104463/air-force-research-laboratory/
134 Britt Theorin, 'The HAARP Project and Non-lethal Weapons. Experts Alarmed—Public Debate Needed,' European Parliament, February 9, 1998, available at: www.europarl.europa.eu/press/sdp/backg/en/1998/b980209.htm
135 Quoted from: Sharon Weinberger, 'Russian Journal: HAARP Could Capsize Planet,' *Wired*, January 8, 2008, available at: www.wired.com/2008/01/russian-journal/
136 Peter Dockrill, 'China and Russia Have Run Controversial Experiments That Modified the Earth's Atmosphere,' *Science Alert*, December 19, 2018, available at: www.sciencealert.com/china-and-russia-conducted-controversial-experiments-that-modified-earth-s-atmosphere
137 Stephen Chen, 'China and Russia Band Together on Controversial Heating Experiments to Modify the Atmosphere,' *South China Morning Post*, December 17, 2018, available at: www.scmp.com/news/china/science/article/2178214/china-and-russia-band-together-controversial-heating-experiments
138 Chen, 'China and Russia Band Together on Controversial Heating Experiments to Modify the Atmosphere.'
139 Eli Kintsch, *Hack the Planet: Science's Best Hope—or Worst Nightmare—for Averting Climate Catastrophe* (Hoboken, NJ: John Wiley, 2010), 3–12.
140 James Temple, 'What Is Geoengineering—and Why Should You Care?' *MIT Technology Review*, August 19, 2019, available at: www.technologyreview.com/2019/08/09/615/what-is-geoengineering-and-why-should-you-care-climate-change-harvard/
141 Jeff Goodell, *How to Cool the Planet: Geoengineering and the Audacious Quest to Fix Earth's Climate* (Boston, MA: Houghton Mifflin Harcourt, 2010).
142 Tim McDonnell, 'The CIA Is Shuttering Secretive Climate Research Program,' *Mother Jones*, May 21, 2015, available at: www.motherjones.com/environment/2015/05/cia-closing-its-main-climate-research-program/
143 Ian Sample, 'Spy Agencies Fund Climate Research in Hunt for Weather Weapon, Scientist Fears,' *The Guardian*, February 15, 2015, available at: www.theguardian.com/environment/2015/feb/15/spy-agencies-fund-climate-research-weather-weapon-claim

144 John Vidal, 'Could an Artificial Volcano Cool the Planet by Dimming the Sun,' *The Guardian*, February 6, 2012, available at: www.theguardian.com/environment/blog/2012/feb/06/artificial-volcano-cool-planet-sun?INTCMP=SRCH
145 James Temple, 'A First of Its Kind Geoengineering Experiment Is about to Take Place,' *MIT Technology Review*, February 19, 2021, available at: www.technologyreview.com/2021/02/19/1018813/harvard-first-geoengineering-experiments-in-stratosphere-sweden/
146 Kintsch, *Hack the Planet*, 69–70; Laura Paddison, 'The Controversial Climate Solution Could Be Exactly What the Planet Needs. Or It Could Be a Colossal Disaster,' CNN, February 12, 2023, available at: www.cnn.com/2023/02/12/world/solar-dimming-geoengineering-climate-solution-intl/index.html
147 James Temple, 'What Mexico's Planned Geoengineering Restrictions Mean for the Future of the Field,' *MIT Technology Review*, January 20, 2023, available at: www.technologyreview.com/2023/01/20/1067146/what-mexicos-planned-geoengineering-restrictions-mean-for-the-future-of-the-field/

References

Adams, Thomas K., *US Special Operations Forces in Action: The Challenge of Unconventional Warfare* (London: Frank Cass, 1998).

Andrew, Christopher and Mitrokhin, Vasili, *The Mitrokhin Archive: The KGB in Europe and the West* (London: Penguin Books, 1999).

BBC, 'China Inducing Rainfall to Combat Severe Drought,' *BBC News*, August 17, 2022, available at: www.bbc.com/news/world-asia-china-62573547

Beason, Doug, *The E-Bomb: How America's Directed Energy Weapons Will Change the Way Future Wars Will Be Fought* (Cambridge, MA: Da Capo Press, 2005).

Bedard, Paul, 'Dry Run: Balloons Called "Top Delivery Platform" for Nuclear EMP Attack,' *Washington Examiner*, February 3, 2023, available at: www.washingtonexaminer.com/news/washington-secrets/balloons-called-top-delivery-platform-for-nuclear-emp-attack

Bernstein, Barton J., 'Radiological Warfare: A Path Not Taken,' *Bulletin of Atomic Scientists* 41, no. 7 (1985): 44–49.

Bertell, Rosalie, *Planet Earth: The Latest Weapon of War* (Dublin: Talma Studios, 2021).

Blank, Stephen, '"Putin's Asymmetric Strategy": Nuclear and New-Type Weapons in Russian Defense Policy,' in: Glen E. Howard and Matthew Czekaj (eds.), *Russia's Military Strategy and Doctrine* (Washington, DC: Jamestown Foundation Press, 2019), 251–301.

Bourke, Emily, 'Military Archives Show NZ and US Conducted Secret Tsunami Bomb Tests,' Australian Broadcasting Corporation, February 3, 2013.

Britannica Dictionary, 'Sabotage,' available at: www.britannica.com/dictionary/sabotage, accessed July 4, 2023.

Bronk, Christopher and Tikk-Ringas, Eneken, 'The Cyber Attack on Saudi Aramco,' *Survival* 55, no. 2 (2013): 81–96.

Bunn, Matthew, 'The Largest Danger at the Zaporizhzhia Nuclear Power Plant: Intentional Sabotage,' *The Bulletin of Atomic Scientists*, July 6, 2023, available at: https://thebulletin.org/2023/07/the-largest-danger-at-the-zaporizhzhia-nuclear-power-plant-intentional-sabotage/

Bushwick, Sophie, 'Chinese Spy Balloon Has Unexpected Maneuverability,' *Scientific American*, February 3, 2023, available at: www.scientificamerican.com/article/chinese-spy-balloon-has-unexpected-maneuverability/

Chen, Stephen, 'China and Russia Band Together on Controversial Heating Experiments to Modify the Atmosphere,' *South China Morning Post*, December 17, 2018, available at: www.scmp.com/news/china/science/article/2178214/china-and-russia-band-together-controversial-heating-experiments

Clarke, Richard and Knake, Robert, *Cyber War: The Next Threat to National Security and What to Do About It* (New York: Harper Collins, 2010).

Davidson, Helen, 'China "Modified" the Weather to Create Clear Skies for Political Celebration—Study,' *The Guardian*, December 5, 2021, available at: www.theguardian.com/world/2021/dec/06/china-modified-the-weather-to-create-clear-skies-for-political-celebration-study

Day, Dwayne A. and Kennedy, Robert, 'Barbarian in Space: The Secret Space Laser Battle Station of the Cold War,' *The Space Review*, June 5, 2023, available at: www.thespacereview.com/article/4598/1

De Vaal, Alex, *Mass Starvation: The History and Future of Famine* (Cambridge: Polity Press, 2018).

Dockrill, Peter, 'China and Russia Have Run Controversial Experiments That Modified the Earth's Atmosphere,' *Science Alert*, December 19, 2018, available at: www.sciencealert.com/china-and-russia-conducted-controversial-experiments-that-modified-earth-s-atmosphere

Dugan, Patrick, 'Sunsets and Dawns: The End-Game for 5GW and the Human Era,' in: Daniel H. Abbott (ed.), *The Handbook of 5GW* (Ann Arbor, MI: Nimble Books, 2010), 98–109.

Eastlund, Bernard, 'Method and Apparatus for Altering a Region of the Earth's Atmosphere, Ionosphere, and Magnetosphere,' US4686605A, available at: https://patents.google.com/patent/US4686605A/en

EMP Commission, 'Report of the Commission to Assess the Threat to the United States from Electromagnetic Pulse (EMP) Attack,' EMP Commission Executive Report (2004).

EMP Commission, 'Assessing the Threat from EMP Attack: Volume I,' Executive Report (July 2017), available at: https://apps.dtic.mil/sti/pdfs/AD1051492.pdf, 5.

FBI, 'Ransomware Attacks on Agricultural Cooperatives Potentially Timed to Critical Seasons,' Private Industry Notification, April 20, 2022, available at: www.ic3.gov/Media/News/2022/220420-2.pdf

Fedor, Lauren, McCormick, Myles, and Murphy, Hannah, 'Biden Says "Strong Reason" to Believe Pipeline Hackers Are in Russia,' *Financial Times*, May 13, 2021.

Fitzgerald, Mary G., 'The Russian Military's Strategy for "Sixth Generation" Warfare,' *Orbis* (Summer 1994): 457–476.

———, *Impact of the RMA on Russian Military Affairs* (Washington, DC: Hudson Institute, 1997).

Ford, Matt, 'Why Doesn't the Government Stockpile Other Essentials as It Does Oil?' *The New Republic*, November 24, 2021, available at: https://newrepublic.com/article/164537/joe-biden-gas-national-reserve

Goodell, Jeff, *How to Cool the Planet: Geoengineering and the Audacious Quest to Fix Earth's Climate* (Boston, MA: Houghton Mifflin Harcourt, 2010).

Graham, William R., Woolsey, James, and Pry, Peter, 'Prepare for the Worst,' RealClear Defense.com, October 21, 2019, available at: www.realcleardefense.com/articles/2019/10/21/prepare_for_the_worst_114802.html

Guillemin, Jeanne, *Biological Weapons: From the Invention of State-Sponsored Programs to Contemporary Bioterrorism* (New York: Columbia University Press, 2005).

Hamblin, Jacob Darwin, *Arming Mother Nature: The Birth of Catastrophic Environmentalism* (Oxford: Oxford University Press, 2013).

Hartog, Eva, 'Moscow Ready to Use Cloud Clearing Techniques on Victory Day,' *Moscow Times*, April 29, 2014, available at: www.themoscowtimes.com/2014/04/29/moscow-ready-to-use-cloud-clearing-techniques-on-victory-day-a34859

Herring, Adam, 'Searching for 5GW,' in: Daniel H. Abbott (ed.), *The Handbook of 5GW* (Ann Arbor, MI: Nimble Books, 2010), 63–74.

Horn, Bernd, Kiras, James D., and Spencer, Emily (eds.), *The (In)visible Hand: Strategic Sabotage Case Studies* (Ottawa: Canadian Special Forces Command, 2021).

House, Tamzy J., Near, James B., Shields, William B., Celentano, Ronald J., Husband, David M., Mercer, Ann E., and Pugh, James E., 'Weather as a Force Multiplier: Owning the Weather by 2025,' Air Force 2025, August 1996.

Johnson, Loch K., *The Third Option: Covert Action and American Foreign Policy* (Oxford: Oxford University Press, 2022).

Kaplan, Fred, *Dark Territory: The Secret History of Cyber War* (New York: Simon & Schuster, 2016).

Keeter, Hunter, 'Non-Nuclear EMP Device Could Threaten Civil, Military Computer Systems,' *Defense Daily* 210, no. 23 (2001): 1.

Keller, John, 'Air Force Deploys B-52 Missiles That Could Disable Enemy Military Electronics with High-Power Microwaves,' *Military Aerospace Electronics*, May 17, 2019, available at: www.militaryaerospace.com/rf-analog/article/14033453/air-force-deploys-b52-missiles-that-could-disable-enemy-military-electronics-with-highpower-microwaves

Kevany, Sophie, 'Avian Flu Has Led to the Killing of 140 Million Farmed Birds since Last October,' *The Guardian*, December 9, 2022, available at: www.theguardian.com/environment/2022/dec/09/avian-flu-has-led-to-the-killing-of-140m-farmed-birds-since-last-october

Khripunov, Igor, 'The Social and Psychological Impact of Radiological Terrorism,' *Nonproliferation Review* 13, no. 2 (2006): 275–316.

Kintsch, Eli, *Hack the Planet: Science's Best Hope—or Worst Nightmare—for Averting Climate Catastrophe* (Hoboken, NJ: John Wiley, 2010).

Koppel, Ted, *Lights Out: A Cyberattack, a Nation Unprepared Surviving the Aftermath* (New York: Broadway Books, 2015).

Koren, Ore, Bagozzi, Benjamin, and Benson, Thomas, 'Food and Water Insecurity as Causes of Social Unrest: Evidence from Geolocated Twitter Data,' *Journal of Peace Research* 58, no. 1 (2021): 67–82.

Leetaru, Kalev, 'Could Venezuela's Power Outage Really Be a Cyber Attack?' *Forbes*, March 9, 2019, available at: www.forbes.com/sites/kalevleetaru/2019/03/09/could-venezuelas-power-outage-really-be-a-cyber-attack/?sh=8fd7b7c607c7

Leovy, Jill, 'Cyberattack Cost Maersk as Much as $300 Million and Disrupted Operations for Two Weeks,' *Los Angeles Times*, August 17, 2023.

Levitin, Carl, 'Russian Documents Set Out "Tectonic Weapon" Research,' *Nature* 383 (1996): 471.

Libicki, Martin, 'The Specter of Non-obvious Warfare,' *Strategic Studies Quarterly* (Fall 2012): 88–101.

Lindsay, Jon R., 'Stuxnet and the Limits of Cyber Warfare,' *Security Studies* 22, no. 3 (2013): 365–404.

Lu, Yonglong, Yuan, Jingjing, Du, Di, Sun, Bin, and Yi, Xiaojie, 'Monitoring Long-Term Ecological Impacts from Release of Fukushima Radiation into Ocean,' *Geography and Sustainability* 2 (2021): 95–98.

Lunev, Stanislav, *Through the Eyes of the Enemies: Russia's Highest Ranking Military Defector Reveals Why Russia Is More Dangerous than Ever* (Washington, DC: Regnery Publishing, 1998).

Lyngaas, Sean, 'Cyberattack on Food Giant Dole Temporarily Shuts Down North America Production, Company Memo Says,' CNN, February 22, 2023, available at: www.cnn.com/2023/02/22/business/dole-cyberattack/index.html

Makichuk, Dave, 'China's EMP Missile Would Plunge Cities into Darkness,' *Asia Times*, September 29, 2021, available at: https://asiatimes.com/2021/09/chinas-emp-missile-would-plunge-cities-into-darkness/

Malet, David, *Biotechnology and International Security* (Lanham, MD: Rowman & Littlefield, 2016).

Manjikian, Mary, *Introduction to Cyber Politics and Policy* (Los Angeles, CA: Sage, 2021).

Manning, Jeanne and Begich, Nick, *Angels Don't Play This HAARP: Advances in Tesla Technology* (Anchorage, AK: Earthpulse Press, 1995).

Margolin, Josh, Sweeney, Sam, and Owen, Quinn, 'Cyberattacks Reported at US Airports,' *ABC News*, October 10, 2022, available at: https://abcnews.go.com/Technology/cyberattacks-reported-us-airports/story?id=91287965

Masters, Roger D., *Fortune Is a River: Leonardo da Vinci's Magnificent Dream to Change the Course of Florentine History* (New York: Plume, 1999).

McDermott, Roger, 'Russia's Entry to Sixth-Generation Warfare: The "Non-Contact" Experiment in Syria,' Jamestown Foundation, May 29, 2021, available at: https://jamestown.org/program/russias-entry-to-sixth-generation-warfare-the-non-contact-experiment-in-syria/

McDonnell, Tim, 'The CIA Is Shuttering Secretive Climate Research Program,' *Mother Jones*, May 21, 2015, available at: www.motherjones.com/environment/2015/05/cia-closing-its-main-climate-research-program/

Meyer, Samuel, Bidgood, Sarah, and Potter, William C., 'Death Dust: The Little-Known Story of U.S. and Soviet Pursuit of Radiological Weapons,' *International Security* 45, no. 2 (2020): 51–94.

Moore Geist, Edward, 'Would Russia's Undersea "Doomsday Drone" Carry a Cobalt Bomb?' *Bulletin of Atomic Scientists* 72, no. 4 (2016): 238–242.

Muntean, Pete and Wallace, Gregory, 'FAA System Outage Causes Thousands of Flight Delays and Cancellations across the US,' CNN, January 11, 2023, available at: www.cnn.com/travel/article/faa-computer-outage-flights-grounded/index.html

National Academies of Sciences, *Terrorism and the Electric Power Delivery System* (Washington, DC: National Academies Press, 2012).

Newdick, Thomas, 'First Laser Weapon for a Fighter Delivered to the Air Force,' The Verge, July 22, 2022, available at: www.thedrive.com/the-war-zone/first-laser-weapon-for-a-fighter-delivered-to-the-air-force

Paddison, Laura, 'The Controversial Climate Solution Could Be Exactly What the Planet Needs. Or It Could Be a Colossal Disaster,' CNN, February 12, 2023, available at: www.cnn.com/2023/02/12/world/solar-dimming-geoengineering-climate-solution-intl/index.html

Paddock, Richard C., 'Lebed Says Russia Has Lost Track of 100 Nuclear Bombs,' *Los Angeles Times*, September 9, 1997.

Page, Carly, 'Maritime Giant DNV Says 1,000 Ships Affected by Ransomware Attack,' *TechCrunch*, January 18, 2023, available at: https://techcrunch.com/2023/01/18/dnv-norway-shipping-ransomware/

PakDefense, '6th Generation of Warfare: Evolution of Warfare from Classic to Hybrid Domains,' PakDefense.com, March 28. 2023, available at: www.pakdefense.com/

blog/6th-generation-warfare-evolution-of-warfare-from-classic-to-hybrid-domains/, accessed July 4, 2023.

Pampinella, Stephen, 'The Construction of 5GW,' in: Daniel H. Abbott (ed.), *The Handbook of 5GW* (Ann Arbor, MI: Nimble Books, 2010), 47–62.

Petersen, Michael, 'Agroterrorism and Foot-and-Mouth Disease: Is the United States Prepared?' *Counterproliferation Papers Future Warfare Series* 13 (2002).

Pry, Peter Vincent, 'Nuclear EMP Attack Scenarios and Combined-Arms Cyber Warfare,' Report to the Commission to Assess the Threat to the United States from Electromagnetic Pulse (EMP) Attack, July 2017.

———, 'China: EMP Threat,' EMP Task Force on National and Homeland Security, June 10, 2020, available at: https://info.publicintelligence.net/US-ChinaEMPThreat.pdf

———, *Blackout Warfare: Attacking the U.S. Power Grid, a Revolution in Military Affairs* (Washington, DC: EMP Task Force on National and Homeland Security, 2021).

Psaropoulos, John, 'Europe Awakens to the Threat of Sabotage by Russian Agents,' Al Jazeera, January 17, 2023, available at: www.aljazeera.com/news/2023/1/17/europe-awakens-to-the-threat-of-sabotage-by-russian-agents

Reuters, 'Massive Network Outage in Canada Hits Homes, ATMs and 911 Emergency Lines,' *The Guardian*, July 8, 2022, available at: www.theguardian.com/world/2022/jul/08/internet-down-canada-rogers-mobile-network-outage

Reuters, 'Russia Is Seeding Rain Clouds in Siberia to Fight Wildfires That Have Been Burning across the Country for Weeks, Reducing Them to Just One Third of Their Former Size,' *Daily Mail*, July 10, 2020, available at: www.dailymail.co.uk/sciencetech/article-8510107/Russia-seeds-clouds-Siberia-douse-raging-wildfires.html

Rid, Thomas, *Cyber War Will Not Take Place* (Oxford: Oxford University Press, 2013).

Rogoway, Tyler, 'Video Appears to Show China Testing Hypersonic Glide Vehicles via High Altitude Balloon,' *TheDrive.com*, September 21, 2018, available at: www.thedrive.com/the-war-zone/23758/video-appears-to-show-china-testing-hypersonic-glide-vehicles-via-high-altitude-balloon

Salama, Vivian, Youssef, Nancy, and Gordon, Michael R., 'Chinese Spy Balloon Tracked over U.S. This Week; Incident Comes Days before Secretary of State Anthony Blinken's Trip to China,' *Wall Street Journal*, February 3, 2023.

Sample, Ian, 'Spy Agencies Fund Climate Research in Hunt for Weather Weapon, Scientist Fears,' *The Guardian*, February 15, 2015, available at: www.theguardian.com/environment/2015/feb/15/spy-agencies-fund-climate-research-weather-weapon-claim

Sanger, David, Krauss, Clifford, and Perlroth, Nicole, 'Cyberattack Forces Shutdown of a Top U.S. Pipeline,' *New York Times*, May 8, 2021.

Sanger, David and Mazzetti, Mark, 'U.S. Had Cyberattack Plan if Iran Nuclear Dispute Led to Conflict,' *New York Times*, February 16, 2016.

Sanger, David and Perlroth, Nicole, 'U.S. Escalates Online Attacks on Russia's Power Grid,' *New York Times*, June 15, 2019.

Sganga, Nicole, 'Physical Attacks on Power Grid Rose by 71% Last Year, Compared to 2021,' *CBS News*, February 22, 2023, available at: www.cbsnews.com/news/physical-attacks-on-power-grid-rose-by-71-last-year-compared-to-2021/

SIPRI, *Weapons of Mass Destruction and the Environment* (London: Routledge, 1977/2021).

Temple, James, 'What Is Geoengineering—and Why Should You Care?' *MIT Technology Review*, August 19, 2019, available at: www.technologyreview.com/2019/08/09/615/what-is-geoengineering-and-why-should-you-care-climate-change-harvard/

———, 'A First of Its Kind Geoengineering Experiment Is about to Take Place,' *MIT Technology Review*, February 19, 2021, available at: www.technologyreview.com/2021/02/19/1018813/harvard-first-geoengineering-experiments-in-stratosphere-sweden/

———, 'What Mexico's Planned Geoengineering Restrictions Mean for the Future of the Field,' *MIT Technology Review*, January 20, 2023, available at: www.technologyreview.com/2023/01/20/1067146/what-mexicos-planned-geoengineering-restrictions-mean-for-the-future-of-the-field/

Theorin, Britt, 'The HAARP Project and Non-lethal Weapons. Experts Alarmed—Public Debate Needed,' European Parliament, February 9, 1998, available at: www.europarl.europa.eu/press/sdp/backg/en/1998/b980209.htm

Thomas, Timothy, 'Russian Forecasts of Future of War,' *Military Review* (May–June 2019): 84–93.

———, 'Advanced Weaponry and the Russian Art of War,' MITRE Center for Technology and National Security, October 2020, available at: www.mitre.org/sites/default/files/2021-11/prs-20-1890-russian-military-art-and-advanced-weaponry.pdf

Toukan, Abdullah, 'The Gulf War and the Environment: The Need for a Treaty Prohibiting Ecological Destruction as a Weapon of War,' *The Fletcher Forum of World Affairs* 15, no. 2 (1991): 97–98.

Trevithick, Joseph, 'Exposed: The U.S. Army Tested Its Own Nuclear "Dirty Bomb,"' *National Interest*, October 14, 2015, available at: https://nationalinterest.org/blog/the-buzz/exposed-the-us-army-tested-its-own-nuclear-dirty-bombs%E2%80%99-14074

U.S. Congress, 'Weather Modification,' United States Senate, Subcommittee on Oceans and International Environment of the Committee on Foreign Relations, March 20, 1974.

U.S. Energy Information Administration, 'U.S. Energy Facts Explained,' available at: www.eia.gov/energyexplained/us-energy-facts/, last updated June 10, 2022.

Van Puyvelde, Damien and Brantley, Aaron, *Cybersecurity: Politics, Governance and Conflict in Cyberspace* (Cambridge: Polity, 2019).

Vidal, John, 'Could an Artificial Volcano Cool the Planet by Dimming the Sun,' *The Guardian*, February 6, 2012, available at: www.theguardian.com/environment/blog/2012/feb/06/artificial-volcano-cool-planet-sun?INTCMP=SRCH

Vidal, John and Weinstein, Helen, 'RAF Rainmakers "Caused 1952 Flood,"' *The Guardian*, August 30, 2007, available at: www.theguardian.com/uk/2001/aug/30/sillyseason.physicalsciences

Waterworth, Michael D., 'Laser Ignition Device and Its Application to Forestry, Fire and Land Management,' U.S. Department of Agriculture (1987), available at: www.fs.usda.gov/psw/publications/documents/psw_gtr101/psw_gtr101_01_waterworth.pdf

Weinberger, Sharon, 'Russian Journal: HAARP Could Capsize Planet,' *Wired*, January 8, 2008, available at: www.wired.com/2008/01/russian-journal/

———, 'Alaska Military Site That Fueled Conspiracy Theories Will Be Transferred to Civilian Operators,' *The Intercept*, July 13, 2015, available at: https://theintercept.com/2015/07/13/military-site-center-conspiracy-will-transferred-civilian-operators/

Whitby, Simon, Millett, Piers, and Dando, Malcom, 'The Potential for Abuse of Genetics in Militarily Significant Biological Weapons,' *Medicine, Conflict and Survival* 18, no. 2 (2002): 138–156.

WikiLeaks, 'Vault 7: CIA Hacking Tools Revealed,' Press Release, March 7, 2017, available at: https://wikileaks.org/ciav7p1/, accessed July 4, 2023.

World Economic Forum, 'What the COVID-19 Pandemic Teaches Us about Cybersecurity—and How to Prepare for the Inevitable Global Cyberattack,' World Economic Forum,

June 1, 2020, available at: www.weforum.org/agenda/2020/06/covid-19-pandemic-teaches-us-about-cybersecurity-cyberattack-cyber-pandemic-risk-virus/

Xu, Adam, 'US Moves to Pull Chinese Equipment from Its Power Grid,' *Voice of America*, May 9, 2020, available at: www.voanews.com/a/east-asia-pacific_voa-news-china_us-moves-pull-chinese-equipment-its-power-grid/6188992.html

Youssef, Nancy, 'Chinese Balloon Used American Tech to Spy on Americans; Preliminary Findings Show the Craft Collected Photos and Videos but Didn't Appear to Transmit Them, Officials Say,' *Wall Street Journal*, June 29, 2023.

Zetter, Kim, *Countdown to Zero Day: Stuxnet and the Launch of the World's First Digital Weapon* (New York: Crown, 2014).

Index

5G telecommunications 44, 137

Abbott, D. 5–7, 9, 10, 33, 37, 38, 47,
    48–49, 52, 58, 61–62, 89–90, 95, 96,
    153, 181, 188, 189, 191
Agroterrorism 175, 176, 185, 191
Amerithrax 46, 61, 142
Arab Spring 98, 102, 105–107, 115,
    117–118, 122–123, 127
Artificial intelligence (AI) 6, 50, 114,
    129–132, 135, 139, 146–151, 155–158
Asymmetric warfare 11, 27, 160
Atmospheric geoengineering 179–180

Behavior change wheel 86–87
Biotechnology 6, 46, 55, 128–129, 143,
    144, 152–153, 157, 185, 190
Bioterrorism 122, 142, 143, 152, 156, 159,
    175, 185, 188
Bioweapons 111, 115, 121–122, 124, 127;
    genetic bioweapons 143–144
Black Knights 73
Brain-computer interface (BCI) 132

Cambridge Analytica 136
Central Bank Digital Currencies (CBDCs)
    140, 141, 150, 151, 154, 156–159
Censorship 81, 88, 92, 96, 118, 124
China 135; collecting DNA 144; Chinese
    spy balloon 170–171
CIA 4, 18, 50, 58–59, 63–64, 66–67, 79,
    103, 107, 117, 124, 166, 179, 182, 186,
    190, 192
Clausewitz 1–2, 8, 10, 12–13, 20–21, 26,
    38, 40, 59, 66, 72
Climate change 132–133, 148, 157, 179,
    186, 191
Cognitive bias 83
Cold War 4–5, 7–10, 14–16, 29, 58–59, 64,
    75, 117, 124, 172–173, 175, 180, 184, 188

Color revolutions 106–107, 118, 123, 125
Complex systems 52
Conditioning 68, 84–85
Conspiracy theory 4, 8–10, 53, 57, 61, 63,
    65, 81, 98–101, 111, 186, 192
Corruption 4, 56, 72, 87, 105
Counterinsurgency 7, 19, 26, 33–34, 39, 46,
    68–70, 72–73, 88–90, 95–97, 161
Counterterrorism 18, 24, 26, 32, 36, 39
Covert action 4, 18, 50, 54, 62, 64, 117,
    122, 124, 185, 189
COVID-19 94–95, 98, 110, 113–116,
    120–127, 181, 192; lab leak theory 111,
    112; COVID vaccines 131, 148, 156;
    COVID as biowarfare 142; COVID
    testing 144
Critical infrastructure 2, 5, 50, 162–163,
    165–166, 170
Cryptocurrency 135, 140, 150, 159
Cyberattacks 2, 164, 166–169, 172,
    180–183, 189–193
Cyber warfare 38, 42, 44, 50, 150, 155,
    160, 162, 165, 169, 182, 184, 189,
    191

Deep state 51, 56–59, 63, 65–66
Digital twins 138–139, 150, 156
Directed Energy Weapons (DEWs) 98,
    116, 119, 123, 145, 183, 187; SDI 173;
    covert DEW attacks 144
Disinformation 2, 23, 50, 62, 63, 66, 79–81,
    85, 88, 90–92, 94, 96–97, 112–113
Disinformation Governance Board 88

Education 81, 83–84, 87, 92, 93–94, 104,
    133
Electromagnetic Pulse (EMP) 164,
    169–170, 180–181, 183, 184, 187–191;
    nuclear EMP 170–171; non-nuclear
    EMP 171–172

EMP Commission 164, 169–170, 172, 183, 188
Environmental modification 43, 172–173, 175, 180
European peace movement 53

Facebook 105, 118, 122, 125
Failed states 15
Fire (as a weapon of war) 173
Food 15, 80, 86, 141–143, 152, 154, 161, 163–165, 167, 173–175, 180, 182
Fourth generation warfare (4GW) 5–7, 11, 19–30, 33, 34, 38, 46–50, 61, 68, 161

Galeotti, M 3, 8, 9
Gaslighting 83, 92, 94, 97
Genetic engineering 129–130, 132, 140, 147–148, 154, 157; genetically engineered bioweapons 143–144, 152, 153, 156, 160, 175, 185, 192
Globalization 14, 29, 47, 49, 54, 62, 65
Google 104–106, 118, 123, 136, 149, 155, 186, 188
Gradients of war 49–51, 54, 58, 161
Gray zone conflict 2–3, 8, 10, 54, 60, 64, 162
Great Terror 98–99, 116, 124, 128
Guerrilla warfare 13–14, 23, 31, 37, 42, 68, 71–72, 89, 97

Hammes, T. 9–11, 21–23, 25, 28, 30, 32–35, 37–38, 46, 54, 59, 61–62, 65, 142, 152, 156
Havana Syndrome 107–110, 115–120, 123–127
Hoffman, F. 8, 10, 19–20, 32, 35
Human domain 7, 46, 68–71, 73, 89, 97, 102, 107, 110, 115, 128, 161, 163
Human terrain 7, 70–71, 73, 89, 94
Huntington, S.P. 18, 32, 36, 62, 65
Hybrid war 2–3, 8, 10, 19, 20, 32–36, 54

Ideology 6, 7, 18, 47, 50, 53, 55, 75, 77–78, 83, 84, 91, 94, 145
Implants 132, 137, 141, 148, 150, 157–159, 168
Information warfare 4, 5, 9, 43, 85, 91, 96
International law 2–3, 12, 45, 55
Ionospheric heaters 178–179

KGB 53, 63–64, 91, 94, 142, 162, 181, 187

Large Power Transformers 168
Lind, W. 5, 9–11, 18, 20–24, 26, 29–30, 32–34, 36, 47, 61–62, 65

Mao Tse-Tung 22, 42, 82; Cultural Revolution 84, 128
Marxism 77–78, 85, 91, 95
Mass formation 82, 83
McFate, S. 14–15, 17, 23, 31, 33, 36, 57, 63, 65
Microwave weapons 108, 119, 124; high-powered microwave (HPM) weapons 171
Milgram experiment 82, 92, 95–96
Mind uploading 130
Nanotechnology 6, 55, 128–131, 144, 147–148, 155–156

Narratives 25, 44, 72–73, 77, 80, 83, 88, 98, 109, 114
National Endowment for Democracy (NED) 103–105
NATO 5, 8–10, 28–29, 34, 37, 53, 85, 89, 93, 94–95, 162, 170
Neocortical warfare 39–40
Neofeudalism 140
Neomedievalism 14, 31, 35
Network organization 19, 22–25, 32–35, 46, 47, 49, 51, 52–53, 70, 102, 104, 118, 125, 134; netwar 40–41; network warfare 44; Trumanite network 57; Spamouflage Dragon network 113, 121, 125; neural networks 130
Neurostrike weapons 145–146, 153, 155
Neurotechnology 6, 128–129, 131–132
New Physical Principle (NPP) weapons 145, 160
New wars theory 22–23, 27, 30, 36
Nongovernmental organizations (NGOs) 15, 41, 45, 78, 98, 103–104, 107, 115, 118, 123
Non-lethality 38–39, 42–43, 59, 64, 65–66, 145, 186, 192
Noopolitik 40–41, 58, 59–60, 64
Nuclear weapons 1, 8–9, 13, 29, 53–54, 63, 145, 153, 158, 160, 166, 169, 170, 172–174, 177–178, 180–181, 183–184, 187, 190–192
Nudging 85–86

Obsolescence of war 1, 8, 10, 42
OODA loop 9, 47–48, 61, 64
Organized crime 55–56, 63, 66–67

Private armies 12, 16–17, 22–24, 28, 30–31, 36–37
Propaganda 2, 19, 50, 72, 74, 76–79, 81, 83–85, 88, 90–92, 94–97, 101, 103, 105, 117, 134; gray propaganda 78; black

propaganda 80; counterpropaganda
80–81, 91, 96; COVID-19 propaganda
113–114, 121, 127
Psychological warfare 4–6, 44, 68, 73–81,
88, 90–91, 96–97, 145

Radiation (as a weapon of war) 154, 157,
173–175, 185, 189
Reflexive control 85, 93, 95, 97
Revolution in military affairs (RMA) 11,
22, 170, 186, 188
Russia 1, 4, 5, 9, 17, 28–29, 31, 34–37, 43,
56, 58, 60, 63, 66–67, 77, 85, 91, 93, 95,
97, 106–109, 115, 118–119, 123–125,
142, 144–145, 152–153, 156–158,
160–162, 166–172, 175–183, 185–192
Rwandan genocide 16, 145, 153, 157, 159

Sabotage 2, 18, 44, 101, 142, 161–169,
172–175, 181, 185–186, 189, 191
Schmitt, C. 13, 30–31, 37, 57
Secrecy in 5GW 50, 52, 57–58, 85, 99, 111,
142, 152, 155, 177
Sentient World Simulation 138–139, 150,
154
Shadow wars 17, 18, 23, 32, 36, 152, 155
Sharp, G. 103, 117, 126
Silent war 26
Smart cities 139
Social credit 141
Social engineering 6, 68, 81, 83–85, 87–88,
92, 95
Social media 78–79, 81, 86, 88, 91, 97,
102, 104–105, 113–116, 134, 135, 139
Soviet Union 4, 8, 10, 16, 53, 58, 63, 64,
82, 101, 116, 124–125, 177, 182; Soviet
ideological subversion 77–78; Soviet
Union biological weapons 142, 143
Stalin, J. 82, 98–102, 115–117, 124–128
Stasi 53, 54, 134, 146, 148, 153–154,
156–157

Super-empowered individuals 6, 46–47,
54–55, 59
Supra-combinations 45–47, 51, 68, 160
Surveillance 2, 11, 39, 57, 58, 92, 95, 104,
134, 139, 148, 149, 155, 157, 159;
surveillance capitalism 135–136
Swarming 41, 52–53, 98

Tectonic weapons 177–178, 186, 189
Terrorism 2, 13, 14, 22, 30, 32, 35, 43,
46–47, 51, 60–61, 66, 91, 96, 99,
104, 122, 172, 182, 185, 189–190;
bioterrorism 142–143, 152, 156, 159,
188; agroterrorism 175–176, 185, 191
Totalitarianism 89, 92, 94, 139
Transhumanism 128, 129, 131–133,
146–147, 156–159
Trinitarian war 12, 26, 27

Ukraine War 1, 7, 17, 28–31, 34, 35, 36–37,
168–169, 175
Unrestricted warfare 2, 8, 10, 41–42,
44–47, 58, 60–61, 64, 73

Vaccines 61, 111, 122, 126, 131, 142–143,
148, 151, 154, 156
Van Creveld, M. 8, 10–12, 17, 23, 26–27,
30–31, 37–38, 59, 66
Vietnam War 13, 28, 31, 35, 71–72, 90, 94,
96, 108, 173

Weather manipulation 164, 173, 176, 179,
180, 185–186, 188–189, 191–192
Westphalian international system 11–12,
14, 22, 24, 26, 29, 33–34, 48
WikiLeaks 105, 117–118, 124, 127, 166,
182, 192
World Health Organization (WHO) 110,
112, 115–116, 120–121, 127, 151, 159

xGW framework 9, 48, 49, 50, 61, 64, 161

For Product Safety Concerns and Information please contact our EU
representative GPSR@taylorandfrancis.com
Taylor & Francis Verlag GmbH, Kaufingerstraße 24, 80331 München, Germany

www.ingramcontent.com/pod-product-compliance
Lightning Source LLC
Chambersburg PA
CBHW060302220326
41598CB00027B/4199